AF769412

Tourism Demand Forecasting

Tourism Demand Forecasting

Jennifer Raga

SOCIETY Publishing

www.societypublishing.com

Tourism Demand Forecasting

Jennifer Raga

Society Publishing
224 Shoreacres Road
Burlington, ON L7L 2H2
Canada
www.societypublishing.com
Email: orders@arclereducation.com

© 2022 Society Publishing

ISBN: 978-1-77469-144-1 (Hardcover)

This book contains information obtained from highly regarded resources. Reprinted material sources are indicated and copyright remains with the original owners. Copyright for images and other graphics remains with the original owners as indicated. A Wide variety of references are listed. Reasonable efforts have been made to publish reliable data. Authors or Editors or Publishers are not responsible for the accuracy of the information in the published chapters or consequences of their use. The publisher assumes no responsibility for any damage or grievance to the persons or property arising out of the use of any materials, instructions, methods or thoughts in the book. The authors or editors and the publisher have attempted to trace the copyright holders of all material reproduced in this publication and apologize to copyright holders if permission has not been obtained. If any copyright holder has not been acknowledged, please write to us so we may rectify.

Notice: Registered trademark of products or corporate names are used only for explanation and identification without intent of infringement.

Society Publishing publishes wide variety of books and eBooks. For more information about Society Publishing and its products, visit our website at www.societypublishing.com.

ABOUT THE AUTHOR

Jennifer is a Tourism and Hospitality Management Specialist with more than 15 years of experience in the industry, giving academic training to University Students and Local Communities about Tourism, Hospitality and Events Management with Eviatour C.A.. She holds a BA in Tourism Business Management and a Postgraduate Diploma in Hospitality Management awarded from the University of Birmingham UK.

TABLE OF CONTENTS

LIST OF FIGURES

LIST OF ABBREVIATIONS

ADLM	Autoregressive Distributed Lag Model
ADSM	Autoregressive Dispersed Slack Model
AI	Artificial Intelligence
AIDS	Almost Ideal Demand System
ARIMA	Autoregressive Integrated Moving Average
ARMA	Autoregressive Moving Average
BRIC	Brazil, Russia, India, and China
BSM	Basic Structural Model
CBS	Computerized Booking Systems
CEO	Chief Executive Officer
CRS	Computer Reservations Systems
CSM	Casual Structural Model
ESG	Environmental, Social, and Governance
GDP	Gross Domestic Product
GDS	Global Distribution Systems
ICT	Information and Communication Technology
IMF	International Monetary Fund
IoT	Internet of Things
MAPE	Mean Absolute Percentage Error
OECD	Organization for Economic Co-operation and Development
SATC	South Australian Tourism Commission
SCI	Supply Chain Integration
SDGs	Sustainable Development Goals
SMEs	Small and Medium-Sized Enterprises
STSMs	Structural Time Series Models
SUV	Sport Utility Vehicle
SVM	Support Vector Machines
SVR	Support Vector Regression

SWOT	Strengths, Weaknesses, Opportunities, and Threats
TFC	Tourism Forecasting Council
TFC	Tourism Forecasting Committee
TRA	Tourism Research Australia
TSCM	Tourism Supply Chain Management
TTH	Travel, Tourism and Hospitality
TVP	Time-Varying Parameter
UN	United Nations
UNESCO	The United Nations Educational, Scientific and Cultural Organization
UNWTO	The United Nations World Tourism Organization
UNWTO	United Nations
UNWTO	World Tourism Organization
VAR	Vector Autoregressive
WTO	World Tourism Organization
WTO	World Trade Organization
WTTC	World Travel and Tourism Council
WWW	World Wide Web

GLOSSARY

A

Agglomeration – a mass or collection of things; an assemblage

Assimilation – the process of taking in and fully understanding information or ideas

Augmented Reality – is an interactive experience of a real-world environment where the objects that reside in the real world are enhanced by computer-generated perceptual information, sometimes across multiple sensory modalities, including visual, auditory, haptic, somatosensory and olfactory

B

Backpacker – a person who travels or hikes carrying their belongings in a rucksack

Ballet – an artistic dance form performed to music, using precise and highly formalized set steps and gestures

Battlefields – or battleground, or field of battle is the location of a present or historic battle involving ground warfare

C

Cadre – a small group of people specially trained for a particular purpose or profession

Cloud Computing – is the on-demand availability of computer system resources, especially data storage and computing power, without direct active management by the user

Commoditization – the action or process of treating something as a mere commodity

Conjecture – a conjecture is a conclusion or a proposition which is suspected to be true due to preliminary supporting evidence, but for which no proof or disproof has yet been found

Cumulative – increasing or increased in quantity, degree, or force by successive additions

Customary – according to the customs or usual practices associated with a particular society, place, or set of circumstances

D

Demography – is the statistical study of populations, especially human beings

Disarray – a state of disorganization or untidiness

Dissemination – the action or fact of spreading something, especially information, widely

Distribution – the act of giving or transporting something to a number of people or places

E

Econometric – is the application of statistical methods to economic data in order to give empirical content to economic relationships

Economy – is an area of the production, distribution and trade, as well as consumption of goods and services by different agents

Ecotourism – tourists experience the natural environment without damaging and disturbing its natural habitat

Emancipate – set free, especially from legal, social, or political restrictions

Enthusiasm – a strong feeling of excitement or interest in something and a desire to become involved in it

Extravagance – lack of restraint in spending money or using resources

F

Feasibility Study – a feasibility study is an assessment of the practicality of a proposed project or system

Flexibilization – refers to the changing work practices by which firms no longer use internal labor markets or implicitly promise employees lifetime job security, but rather seek flexible employment relations that permit them to increase or diminish their workforce, and reassign and redeploy employees with ease

Fling – throw or hurl forcefully

Forecasting – is the process of making predictions based on past and present data and most commonly by analysis of trends

G

GDP – is a monetary measure of the market value of all the final goods and services produced in a specific time period

Geotagging – or GeoTagging, is the process of adding geographical identification metadata to various media such as a geotagged photograph or video, websites, SMS messages, QR Codes or RSS feeds and is a form of geospatial metadata

H

Harmonizing – add notes to (a melody) to produce harmony

Homogenization – a process by which the fat droplets from milk are emulsified and the cream does not separate

I

Immersion – the action of immersing someone or something in a liquid

Impediments – a hindrance or obstruction in doing something

Inbound – traveling towards a particular place, especially when returning to the original point departure

Inertia – a tendency to do nothing or to remain unchanged

Intangible – unable to be touched; not having physical presence

Intraregional – occurring between, or existing between two or more regions interregional communication/trade
Inventory – it refers to the goods and materials that a business holds for the ultimate goal of resale, production, or utilization
Investors – a person or organization that puts money into financial schemes, property, etc. with the expectation of achieving a profit

J
Jeopardized – put (someone or something) into a situation in which there is a danger of loss, harm, or failure

L
Leisure – time when one is not working or occupied; free time

M
Multivariate – means involving multiple dependent variables resulting in one outcome

N
Neoliberal – is a term used to describe the 20th-century resurgence of 19th-century ideas associated with economic liberalism and free-market capitalism
Neural Networks – are a series of algorithms that mimic the operations of a human brain to recognize relationships between vast amounts of data
Novelty – the quality of being new, original, or unusual

O
Optimistic – hopeful and confident about the future
Orbital Rocket – is a spaceflight in which a spacecraft is placed on a trajectory where it could remain in space for at least one orbit
Outbound – traveling away from a particular place, especially on the first leg of the return journey
Outsourcing – it is the business practice of hiring a party outside a company to perform services and create goods that traditionally were performed in-house by the company's own employees and staff

P
Pragmatic – dealing with things sensibly and realistically in a way that is based on practical rather than theoretical considerations

R
Reassessment – the consideration or assessment of something again, in the light of new or different factors
Recreation – it is an activity of leisure being discretionary time
Regression Models – is a form of predictive modelling technique which investigates the relationship between a dependent (target) and independent variable (s) (predictor)

S

Scenic – providing or relating to views of impressive or beautiful natural scenery

Sourcing – obtain from a particular source

Souvenirs – a thing that is kept as a reminder of a person, place, or event

Stakeholders – is a member of "groups without whose support the organization would cease to exist," as defined in the first usage of the word in a 1963 internal memorandum at the Stanford Research Institute

Strategy – it is a general plan to achieve one or more long-term or overall goals under conditions of uncertainty

Suborbital Rocket – refers to vehicles that travel high enough to reach the "edge" of outer space but do not have the energy to achieve orbit

Supply Chain – a supply chain is the network of all the individuals, organizations, resources, activities, and technology involved in the creation and sale of a product

Supply Chain Integration – it is a process where all the parties involved with the fulfilment of a product are integrated into a single system

Sustainable Development – The development which meets the needs of the present without compromising the ability of future generations to meet their own needs

SWOT Analysis – is a strategic planning technique used to help a person or organization identify strengths, weaknesses, opportunities, and threats related to business competition or project planning

T

Thereon – on or following from the thing just mentioned

Traditionalists – an advocate of maintaining tradition, especially so as to resist change

Transportation – it is the movement of humans, animals, and goods from one location to another

V

Voluntourism – a form of tourism in which travelers participate in voluntary work, typically for a charity

PREFACE

This book takes the readers through several different stages of tourism and forecasting necessitated by the tourism sector. This book sheds light on several aspects of tourism demand forecasting, such as creativity and innovations in the tourism business, strategic management in tourism, parameters and techniques used in tourism demand forecasting, challenges and opportunities, and tourism supply chain management. The first stresses on the basic introduction to tourism so that the readers are clear about the philosophies behind the concept as they form the basics in the field. This chapter therefore emphasizes on the classification of tourism, nature of tourism, the importance of tourism, impacts of tourism, potential growth in the tourism sector.

The second chapter takes the readers through the concepts of tourism demand forecasting. This chapter will provide highlights on various aspects such as types of demand forecasting, strategies for demand forecasting, forecasting for tourism and its products, and social and safety factors. Then, the third chapter explains the creativity and innovations in the tourism business. It therefore explains the key technology trends emerging in the travel industry, such as voice search and voice control, robots, contactless payments, and virtual reality.

The fourth chapter introduces the readers to the significance of strategic management in tourism. It therefore explains the various components of the strategic management process, essential tasks of strategic management, principles to the successful strategy management, and the importance of strategic management during a hospitality business. The fifth chapter throws light on various parameters and techniques that have been used in demand forecasting in the sector of tourism. This chapter further contains different approaches to measure tourism and forecasting tourist arrivals using time. It therefore includes varying parameters in structural time series models, and an integrative approach to tourism forecasting.

The sixth chapter is about the challenges and opportunities that are faced in tourism forecasting. The readers are then explained about the special importance of sustainable tourism poverty alleviation in third world nations. The seventh chapter explains the role of supply chain management in the tourism industry. This chapter also emphasizes the sourcing strategies of supply chain management in tourism, components of tourism supply chain management, and the need for tourism supply chain management.

The last chapter of this book sheds light on the new business environment and trends in tourism. This chapter therefore mentions the role of digital technology in tourism, new trends in tourism, and how technology is enabling the digital tourism economy.

This book has been designed to suit the knowledge and pursuit of researchers and scholars and to empower them with various aspects of tourism, and also to help them understand why tourism demand forecasting, so that they are updated with the information. I hope that the readers find the book explanatory and insightful and that this book is referred by scholars across various fields.

Chapter 1

An Introduction to Tourism

CONTENTS

When mankind evolved, the history of travel and tourism came into existence. In order to seek food and shelter, and to explore and find an environment which is friendly and suitable for a better habitation, generally for a specific period of time, people traveled. Towards the beginning of the 15th century, people started shifting to other places and discovering new places for settling according to the favorable climate and environment. Moreover, in the later years, travel became easier due to the development of transport and various measures taken on safety issues. It was influenced by the need to survive. Gradually trade and commerce started, the market and economy flourished. In Italy and Europe, a cultural and intellectual period began to evolve, which is called Renaissance.

It was the rebirth of literature and art in Europe, since at that time, traveling upsurged with the rise in the Industrial Revolution in the 18th century. People migrated to areas which were being developed in order to attain facilities. During the beginning of the 19th century, people started traveling for pleasure. Moreover, when people started traveling for leisure and pleasure, they started visiting museums, cathedrals, monuments. Therefore, people stayed in hotels for as short as two days and as long as six months. Thus, this travel gradually transformed into tourism.

It gradually became a good revenue opportunity as Governments of different countries started earning through tickets on visits to monuments, museums, and places of historical significance. This money from tourism contributed to the economic development of the country. Therefore, tourism grew more and moreover the period, with the development in technology and transport. Moreover, the marketing of the tour operators and travel agencies expanded, which also resulted in the growth of tourism.

1.1. INTRODUCTION

With the evolution of years, the travel and tourism developed significantly. It contributed to the growth of social and economic factors. Through the development of technology, jet travel and communication, travel and tourism became one of the world's largest and fastest growing industries.

Travel and tourism industry, in fact, became a supreme economic force. It covered 12% of the total world trade at 8% annual rates. For regional development as well, tourism has played a vital role, as it helps in reviving new economic interest. It has a beneficial influence on the revenues, income, payments, and production. On the other hand, it may also have a negative impact, specifically on the environment. Tourism expands economic growth but the unexpected and unmanageable growth in tourism results in the decline of the environment due to which the tourist growth sometimes is jeopardized. For instance, many people going on trekking are deteriorating the environment by throwing and dumping waste products here and there, as a result trek for certain places has been banned for definite intervals of time as it has adverse impact on climate and environment.

For the future development in tourism, the environment must be secured and preserved as it is a key factor. People would not want to visit places that are environmentally bad. When one starts to protect the environment, the growth in tourism increases and therefore it contributes to economic development of the nation. This is applicable in terms of tourism which is established on the grounds of natural environment and on historical and cultural heritage.

For the continued growth of tourism, there are three interdependent factors which include:

- economic;
- socio-cultural;
- environmental.

Talking about the sustainable development of tourism, there must be a persistent effort towards ideal use of resources, along with biological diversity, reduction in ecological, social and cultural influences and increase in conservation and local communities' benefits. Proper management should be in place to attain the objectives of sustainable tourism and economic development. Therefore, strategies and policies must be made in order to maintain a healthy environment and climate so that it may attract tourists and increase tourism.

The schemes and programs on the development of tourism, have their impact on the overall economic growth of a country through generating incomes and taxes, economic reconstruction and diversification. Tourism is therefore carried on as an economic task, as it boosts and supports the economy and enhances regional development. Moreover, the revenue generated by the tourists who come to visit different countries becomes the primary source of foreign exchange earnings and fiscal revenue for the government who acts as a host. Hence, the redistribution of national income is done between the markets that have tourist attention and are tourist receiving destinations. They become human capital, by the expenditures they do in a country while traveling. As a result, the economic impacts of tourism on the host destination and region along with the rapid economic changes result into economic development.

Figure 1.1. Tourism plays an important in a country's economy.

Source: Image by unsplash.com

The future development of tourism refers to the sustainable development that fulfills the development in present by using resources in such a manner that they will be available in future for the later generations to use and meet their needs without consuming all the resources, leaving nothing for the future generations. Moreover, the historical monuments, statues, national heritages built during the ancient times and nowadays being used as the tourist attraction should be given proper care and time to time renovations must be done in order to preserve it for a very long period along with the protection of the environment and climate.

It could be said that sustainable development is "the idea of limitations imposed by the state of technology and social organization on the environment's ability to meet present and future needs" (Tay Yu Shan, 2016). Therefore, principles and policies must be set in a framework where all over the globe, the growth and development should be done in such a way that fulfills the needs of society and of the environment of the present as well as of the coming generation. Moreover, schemes and plans must be set for the utilization of resources in an effective and efficient way which also works in the environmental management to maintain economic development along with the upgradation of the environment. Hence, sustainable environmental development is necessary for nourishing the future of tourism and taking it to the peak, so that eventually it may result in economic development all over the globe.

1.2. SUSTAINABLE TOURISM DEVELOPMENT

Along with the concept of sustainable development, studies have underlined and paid attention to the tourism ecology correlation, by also pointing at the harmful impacts of mass tourism on the natural resources, or the man made and socio- cultural resources of host communities. Therefore, the primary objectives of economic welfare should be aimed at considering the protection of natural, social and cultural resources and capital. However, while focusing on the development of tourism, social and economic aspects should never be taken for granted. The idea of sustainable tourism should be viewed as a balanced development that fulfills the requirements and needs of tourists, public and private tourism industry operators which are taken as supplies and the perseverance of natural and cultural resources which acts as a fundamental base for tourism.

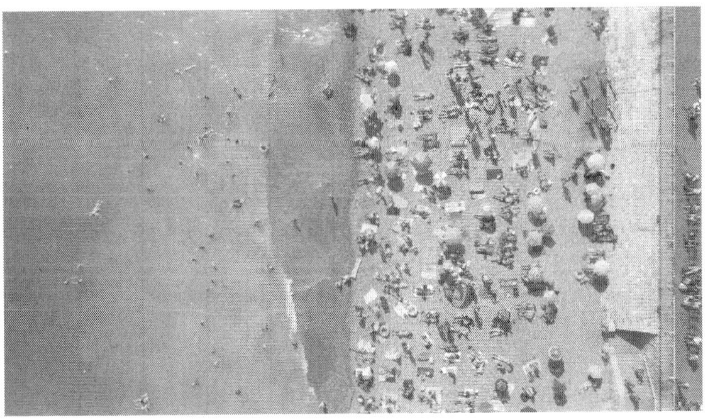

Figure 1.2. Sustainable tourism development.

Source: Image by unsplash.com.

The sectors of tourism should imply tourist policies, and the costs imposed on this sector. All these schemes and policies come with hurdles. There is a socioeconomic impact on the national and regional economy. According to Pearce (1991) they are as follows:

- Balance of payments: In many countries, while the tourism industry is the primary source of foreign exchange earnings, there are also reductions in the net benefits of the balance of payments which are dependent on the actions of the foreign tourist operators.

- Regional development: All over the globe, the tourism industry has grown tremendously, with each region offering its own specialties.
- Diversification of the economy: Economic development is possible due to the multidisciplinary nature of the tourism industry.
- Income levels: The money generated from tourism helps in socio economic development as people get employment at different parts of the industry value chain.
- State revenue: the revenues earned by the states generated from the taxes earned by the public, on their visit. Along with the revenues and taxes there is a requirement of time-to-time maintenance and construction of buildings and monuments due to which expenditures are done.
- Opportunities of employment: When it comes to employment opportunities, there are various countries where tourism is an important source of employment as this industry provides jobs in many sectors. It is also beneficial for the unskilled and semi-unskilled labor force.

There are several principles responsible for the growth and sustainable development of tourism that act as a bridge between socio cultural and environmental aspects. They are as follows:

- Using resources sustainably so that future generations could witness and enjoy the fruits
- Reduction in the overuse of resources and products and preserving it instead of wasting.
- There should be diversification that needs to be maintained.
- Incorporation of tourism into planning.
- Local economies must be backed up and assisted.
- There should be indulgence of local economies.
- There should be involvement of stakeholders as well as of public.
- Staffs who are associated with tourism must be trained.
- Marketing for tourism and research work must be done responsibly.

The aspects and principles incorporated in the development of tourism varies from country to country depending upon the extensive framework of circumstances like the tourism life cycle, promotion strategies for local tourists, use of adequate information systems and marketing schemes in tourism. The nature of tourism is multi activity and multi sectoral, because of which tourist products show a clear difference with the traditional private goods model.

Tourist market includes various types of goods, public goods, mixed goods, private goods with some jointness characteristics and these combinations of goods in the market cannot be enveloped by a market system. Therefore, the suitable measures for a sound economic evaluation of the tourism welfare will not be attainable for most schemes and policies. These effects are obvious and evident, and they must be examined in the light of the tourism profession on the natural, cultural and socio-economic environments of tourism destinations. Moreover, the excessive consumption of resources, pollution and waste caused by the development of tourism infrastructure and facilities, transport and tourism activities, gives rise to the negative effects on the environment.

These adverse impacts are irreversible and uncertain. Most of the time, tourists are not charged by the social costs and the marketed goods with prices per unit are not involved either. This becomes accurate while considering tourism which is strongly based on environment as well as on historical and cultural heritage [Russo-Vander Borg, 2000]. However, the unplanned and immoderate tourism which has its focus on short term satisfaction and interest has an adverse and lasting impact on the environment and on societies. Moreover, tourism which is established on wrecks puts a challenge on whether the development and growth of tourism is viable in a certain region, so that there should not be any superficial negative diseconomies that eventually look up towards the development that should be ecologically sustainable in the long run. Therefore, whether this development is economically possible, it is morally and socially fair for local communities or not. Hence, for the sustainable development of tourism, the relationship between environmental and cultural conservation and tourism growth is necessary.

There are many definitions and interpretations of sustainable development which are applied to many contexts including tourism. Moreover, an accurate definition would be that widely used by the Brundtland Report, 1987: "Sustainable development is one that meets the needs of the present generation without compromising the ability for future generations to meet their own needs."

> The actual meaning of sustainable development remains hidden sometimes, especially when it comes to tourism. Although the concept of sustainable development is a global concept yet there is a failure to acquire and work upon the real meaning of this term.

This includes a comprehensive approach to planning and strategies:

- Preserving and protecting the environment and the man-made heritage/monuments.
- Protection of essential ecological processes.
- Encouraging and engaging public participation
- Making certain of the productivity should be assisted in the long run future.
- Gives an opportunity and assistance across the globe between various countries.

There are certain organizations that are connected and work for the sustainable development in tourism such as the World Tourism Organization (UNWTO) and Tourism Council (WTTC). These organizations work for the upliftment of the tourism industry by following the principles of sustainable development that focuses on keeping the environment clean and improved. For example, the reduction in the damages caused by local people, tourists and the tourism industry. Along with the Earth council they enhance and encourage the industry to take initiatives in protecting the environment in the area they function.

Nevertheless, sustainable development in tourism includes all activities, management and development of tourism that ensures and gives protection to the natural, social and economic integrity along with the conservation of natural and cultural resources. These frameworks of rules and guidelines are relevant in all types of tourism and destinations, in addition to mass tourism and the different functions in the tourism section. Hence, the strategies and schemes used in the viable development of the tourism industry must include making ideal use of environmental resources that is a key factor in the tourism development, with the balance in ecological processes, along with preserving the natural heritage of a country and biodiversity.

However, the socio-cultural authenticity of host communities must be respected and conserved, the cultural heritage and traditional values must be preserved and should contribute to intercultural understanding and tolerance. Possible and useful economic functions which provide long-term socio-economic benefits and provide income earning opportunities and social services to host communities and exercising some relief in poverty (UNEP 2004).

In addition to maintaining high levels of tourist satisfaction, a significant and happy experience to the tourists must also be bestowed, by making them aware of sustainability and promoting sustainable tourism practices amongst them. Sustainable tourism has relevant aspects like competitive and socially responsible tourism business, it also provides chances to take part in tourism activities by giving employment opportunities, satisfaction and advantages to the local communities.

Ample efforts have been made to execute the principles and strategies of sustainable development in tourism at regional as well as tourism enterprises level. Another important dimension of sustainable tourism is the "eco-tourism" which as per The International Ecotourism Society is traveling responsibly to the natural areas that protect the environment and climate at every cost and serve to the welfare of local people and their living. It incorporates the features of rural as well as cultural tourism.

Nevertheless, ecotourism holds certain principles: it works towards the betterment and conservation of natural and cultural heritage, it requires the involvement of local communities when it comes to planning, and tasks assigned for the growth and development. Therefore, the clear reports of description for visitors which gives them proper clarification with respect to the natural and cultural resources. It revolves around solo travelers or sometimes to the small groups of travelers. Moreover, some of the activities and tasks included in eco-tourism are hiking, mountain climbing, rafting, bicycling, trekking, snorkeling, travel to places where flora, fauna and cultural heritages are key attractions. Hence, it is beneficial for the education trips, personal growth of an individual, to make people aware of the importance and beauty of nature, and cultures across the countries, in order to acquire peace of mind by practicing meditation and self-reflection.

Another important characteristic feature of ecotourism is to cater the needs and acquire benefits for the local communities. This can be attained by implementing local staff in various activities, making use of local products, involvement of local people in decision making tasks, and proper management of tourist activities (Sambotin, Patrascoiu, Coroian, Merce, 2011).

1.3. CLASSIFICATION OF TOURISM

Figure 1.3. Classification of tourism.

Source: Image by unsplash.com.

1.3.1. Recreational Tourism

Recreational tourism is also referred to as leisure tourism, which people do in order to take a person out of a monotonous and dull life. To detox and get away from everyday life, people go to spend time at mountains, sea beaches, deserts. The fun, recreation of people is the basic concept of travel and tourism. Most people go to various places away from home in order to spend their vacations.

This has become a worldwide phenomenon because of the development of transport and production, and buying power of consumers. It works upon the mental health of people by keeping them refreshed and stress free from work and daily routine, it lets them enjoy seaside, beaches, hills, exotic views and also encourages them to participate in cultural programs along with the different forms of entertainment.

1.3.2. Cultural Tourism

Cultural tourism focuses on cultural and intellectual curiosity. It satisfies and makes people aware of the cultural knowledge by their visits to ancient monuments, historical and religious places, and cultural heritages. It allows the tourists to take part in cultural activities like a festival or rituals of certain regions. This way they get an opportunity to interact with the local people and be a part of the local community.

And these communities accept their culture as a showcase of cultural tourism which helps in cultural growth. Communities promote their cultures to make it unique for tourist attraction. There are also some benefits of cultural tourism like it has economic impact which is positive, it helps the undeveloped communities with better tourist attraction to establish.

Secondly, social service and development of these local communities can be done properly by the revenue generated from the tourism. Thirdly, it enhances the local goods and markets which flourishes local economies.

1.3.3. Sports and Adventure Tourism

It refers to the activities and adventures which people seek to have for experiencing something for instance, skiing, hiking, trekking, rafting. In mass tourism it results in 4% annual growth rate over the past years. Adventure tourism is increasing day by day which eventually will make it a fastest growing sector in the tourism industry. Travelers who go for adventure tourism are characterized as: Confident, well-traveled consumers who not only enjoy nature and sightseeing but also want something phenomenal from their lives.

They are always wanting for more fun and are authentic, they want all the worldly experiences. They have a very good and thorough knowledge of the environment, different places and of culture and society. They are enthusiastic and active beings born with brave hearts for adventures.

Figure 1.4. Sports and adventure tourism.

Source: Image by unsplash.com.

1.3.4. Health Tourism

When people travel for obtaining the medical facilities and cures for their treatment either its physical health or mental, for instance, spa or yoga, it is called health tourism. When it comes to the national development of tourism, health tourism has always been an important factor. It contributes to the high rates of expenditure to the tourist trip.

People who are sick or those who accompany sick people are the ones who are involved in health tourism. There are multiple options for the people seeking healthcare facilities, they can choose among those options. The health and medical services are provided to them at lower and affordable prices. These travelers or operators reach out to the institutions having international quality and quantity.

They look up for the newest technology in medicines along with the accessibility of personal aid, meaning to say the availability of an expert staff must be provided to the tourists. They must be given pre-defined dates so that they should arrange their travel accordingly without having to wait.

1.3.5. Convention Tourism

Convention tourism is often referred to as business tourism. It is a significant aspect of travel. It refers to the traveling which involves people going overseas for their business meetings or profession related work. It is staying away from the regular environment and traveling abroad for attending conferences, exhibitions, or doing inspections.

It also occupies restaurants or hotels for meetings or dining out with somebody for business related purposes. It has business destinations which are already set and booked, mainly in industrial regions. Convention tourism involves individuals or small groups for attending larger meetings.

1.3.6. Incentive Tourism

The tours set up by different organizations in order to give their employees benefits after achieving certain targets or the trips or the holiday trips granted by the companies to agents or dealers after they obtain high targets are the incentive tourism. This is a new term in the tourism industry which is enlarging day by day. In recent times, incentive tourism in the USA contributes to the business of three billion dollar.

1.4. NATURE OF TOURISM

Tourism is connected to people and traveling. From the time civilization began, humans are associated with the idea of moving and relocating in the quest for a better environment and place. They have a habit of exploring, and therefore knowing the unknown. The nomadic behavior of humans gave rise to the phenomenon of traveling. Moreover, when the industrial revolution came into effect it influenced traveling more to different destinations. It enhanced and improved transport and security facilities due to which traveling increased more. It thus made the living conditions of travelers easier and safer and amplified tourist activities. The nature of tourism is socio economic. It incorporates the experiences of tourists and visitors.

These tourists are backed by the travel and tourism industry and the host destination and the aggregate of these activities are the tourism product. These activities are demonstrated as supply and demand and the strategies made up for tourism development must make a balance between demand and supply. This therefore meets the needs of the market along with the better understanding of market characteristics. Moreover, tourism is a tertiary industry; it provides services for free of cost. The tourism products are produced, kept reserved, sold and eventually consumed. The nature of tourism is intangible, perishable, inseparable and heterogenous. For a sustainable development of tourism, these factors must be worked upon properly. However, the intangibility of tourism is evident by the fact that people are sold on an experience, which they can only feel instead of seeing or touching. They are perishable in a way that the services they provide are not stocked up for later sale or use.

While talking about the inseparability nature of tourism, the producer and the consumer that are the Bahamas and the tourists must come along for the consumption of the tourism activities and products. The factor of production and consumption is done at the same time. The experience is therefore affected by the provider of the tourism and product of the client. Moreover, the heterogeneity of tourism is relevant because there are various people involved who give an anticipated level of quality services in order to make the experience beneficial and sustaining. Hence, the differences in tourism services happen as it is since who provides them, and when, where or how that they are imparted.

Figure 1.5. Nature of tourism.

Source: Image by Simply Writer.

On the other hand, for a well-planned tourism, it is important to focus on the dimensions of supply, which include different modes of transportation available, overall infrastructure related to tourism (hotels, resorts, restaurants, etc.). The information and details given to tourists must be accurate. Therefore, marketing and promotion strategies must not be weak. Hence, the communities that are within the tourist destination need to be maintained from time to time, the political and institutional structure for empowering tourism needs to be set in a proper order.

The factors of demand and supply should be interconnected by the flow of resources like capital, labor, goods and services, tourist expenditures, the flows of marketing and promotion, the artifacts related to tourists and experiences from the destinations into the tourist created regions and areas. Therefore, transportation and security works as an essential bridge between both to and from the destination.

The nature of tourism is also dynamic and complex because of the various elements linked together to it and due to the different sectors playing a vital role in the success of the industry. These aspects and dimensions are connected to facilitate the tourist experience and generation of tourism revenue and markets. Therefore, the dynamicity of tourism constructs the internal and external environment of the destinations on a systematic basis. Moreover, it produces transformations whenever required and provide a healthy, manageable, and operational tourism industry.

This clearly shows that the development of the tourism industry requires various forces and factors, and it can never work on its own, without the involvement of the environment, climate and local communities, the economic, social and cultural factors play a vital role in the construction and growth of the tourism industry.

1.5. IMPORTANCE OF TOURISM

Tourism plays a vital role in the growth and success of a country's economy around the globe. It nurtures the revenue of the economy, provides job opportunities, builds a country into something beautiful and worth visiting by proper maintenance. It therefore gives rise to cultural sensibility, and fuels communication between the foreigners and local people. Moreover, the tourism industry provides various kinds of job opportunities. This involves jobs not only in the tourism sectors but also in various other sectors like communication, education, health etc. There are tourists who travel for exploring different cultures and interacting with different people, practicing and gaining knowledge about different traditions. The cultural aspect of the tourism industry plays a significant role in attaining profits by the local markets, malls, shops, restaurants, hotels and stores.

It is always about giving and taking, the government depends for its large amount of profit on revenue generated from the tourism industry, it also spends and invests a large amount of money on the maintenance of the infrastructure of the country in order to attract more people across the borders. Hence, it ensures safety and up to date facilities for the tourists. Moreover, the maximum work done for the country's upliftment is that made for the sake of inviting more and more tourists. For example, the construction of new roads and highways, well developed markets and clean parks, improving public places, making new airports at different areas, the improvement in schools, hospitals, malls, restaurants, shops etc. Hence, this is done for trading and free flow of goods and services, taking safe and innovative measures. This not only involves the tourist's attraction but also it uplifts the standards of living of local people and communities. It bestows an opportunity to them for their economic and educational development.

Figure 1.6. Importance of tourism.

Source: Image by unsplash.com.

On the other hand, the exchange of values, ideas, cultures and traditions between the foreigners, tourists, and local people to a large extent is possible through the tourism industry. The showcase of different traditions and values by exhibitions, conferences and events spreads awareness among the tourists. Moreover, these exhibitions, conferences, events have registration fees help in taking a good amount of revenue from foreigners. The heavy prices on tickets and transports are another source of generating revenue. It gives several opportunities to foreign tourists to learn and develop a sense of cultural enrichment, along with providing many opportunities to the local citizens as well. Hence, an exceptional growth has been documented in the tourism industry since the year (2011). As stated by (UNWTO) UN Tourism Organization, the international tourist arrivals increased by 7% all over the world in 2011. Since the recorded data from 2010, this data was the highest. Countries earning profits from the tourism industry are the United States, Spain and Australia, Mauritius, Malaysia, Singapore. They generate revenues from the international visitors.

Nevertheless, according to Zurab Pololikashvili, Secretary General of the UN World Tourism Organization, "The growth of tourism in recent years confirms that the sector is today one of the most powerful drivers of economic growth and development." This is true and possible because of the affordable transport facilities and technological impacts which eventually lead to the powerful economic growth.

The importance of tourism is significant in different sectors such as follows.

1.5.1. Employment and Income

As stated earlier the tourism industry creates a lot of job opportunities for people by employing people in various sectors related to traveling like, hotel, restaurants, tour operators, guide etc. It gives jobs in the sectors like hotel and hospitality industry, service, entertainment and transportation industry. Moreover, for generating public and private sectors income tourism is an authentic source. Government revenue is produced from the taxes it charges. The income which comes out of these taxes is the public income. The profits generated by the sellers who sell products in the local market and tourists buy these goods and services is the private income.

1.5.1. Maintenance in Infrastructure

The tourist destination places are maintained and repaired from time to make it more attractive and to get a different kind of status and standard. People wish to go there, to enjoy and make memories because of the aura it holds. Moreover, not only the mending of destinations and communities is done but also it focuses on the construction of dams, roads, airports, railways and every possible connectivity that plays an important role to help tourists visiting different places.

1.5.2. Educational Sectors

Educational tours, courses that need studies in different countries for a short or long period of time, educational exchange programs, paper presentations in different countries, gives rise and upliftment to international tourism. It provides preferable and thorough knowledge about the host countries. People go on education tours to have education or training; it can result for a short or long time period. Therefore, a lot of people move to other countries or regions or travel to foreign countries to obtain education. Importance and awareness of education is gradually increasing so people from rural areas are also shifting to urban areas and enrolling themselves in universities. Sometimes people travel to different regions and countries for a conference, training or courses, this trip is not permanent. It lasts for a week or month.

1.5.3. Amending the Environment

For the conservation of the environment, tourism has taken important steps and set notable goals. It has worked upon the protection of the environment and climate in the regions where historical sites and national heritages are built. Moreover, the preservation of endangered species has also been given importance. It has hence looked after the maintenance of the natural environment by taking every possible step to avoid its damage.

Figure 1.7. Amending the environment.

Source: Image by unsplash.com.

On the other hand, measures like a ban on throwing plastic materials on hills and in water bodies are taken to preserve wildlife and plant species. It can be safely said that such measures are taken by most countries, though the extent to which they are enforced could differ. Tourism industry being the most complex one can have negative and positive effects on the environment, it is the duty of the government of every country to preserve the environment at every cost. Hence, if the surroundings are clean, national heritages are not damaged, weather is fine, beauty and infrastructure of that place is well maintained, it will attract people without any doubt, people seek peace at such places.

1.6. IMPACTS OF TOURISM

In order to maintain, establish and develop, the industry of tourism requires enough money for expenditure, earning and making profits. Tourism has positive as well as negative effects especially on tourist destinations. The impacts of tourism are on the economic, socio cultural, and on environmental aspects. Tourism has negative and positive health consequences on the lives of local people. Hence, by elucidating the positive effects of tourism, there are certain points which must be kept in light.

1.6.1. Positive Impacts

- It creates employment opportunities for various sections of society.
- Enhances local communities, their cultures and protects their traditional values.
- Local communities are built, and the infrastructure is nurtured by the income which generates from the tourist activities.
- Foreign currencies are utilized by local people.
- Preservation of natural environment and heritages through the income from tourism.
- Revenue is generated from foreign countries.
- It gives various opportunities to modest business.

Figure 1.8. Impacts of tourism.

Source: Image by unsplash.com.

1.6.2. Negative Impacts

Negative impacts are as follows:

- The improvement in transportation facilities has a negative effect on the environment as the rise in air travel has led to an increase in carbon dioxide emissions. However, the activities on a local level have proven to be a menace to the natural landscape.
- Although it provides job opportunities to people, yet the local people are employed in the low skill, less wages and miserable working conditions.
- Travel agents and agencies, big hotels earn profit from tourism whereas local guest houses are not profited much.
- When people from different countries come to local communities it sometimes results in the degradation of local cultures.
- At times there is a deterioration of the economy due to the outbound tourism.
- Different types of tourism have different ways of reliance on customers.
- It encourages collateral economies.

Dealing with the negative impacts, proves to be the biggest challenge faced by the tourism industry. The impacts of tourism are precisely on four different aspects. As mentioned above, they are- economic, socio-cultural, political and environmental. These are the factors which enhance sustainability in the tourism industry and are called the pillars of sustainable tourism. Therefore, in order to maintain this sustainability these factors need to be preserved and protected for the further growth and development in tourism. Along with them, the local and natural resources should be conserved as they also play a vital role in sustainable development. Moreover, tourism should be developed in a way that it does not affect or degrade the surroundings, environment or climate or it will never be a sustainable development. In addition to this, the tidiness and neatness of the tourist destination, its beauty must be maintained, which seems appealing to the tourists. It becomes a challenge to create a balance between all these four pillars of sustainable tourist development.

1.6.3. Impacts of Tourism on Economy and Development

There are many people who are involved in the tourism industry, who are associated with the generation of income either directly or indirectly. The sectors which tourism covers when it comes to the economy and development are transportation, accommodation, food, ticketing, guiding, boating, rafting, trekking, maintenance of the tourist destinations. Moreover, when a person comes to visit a country, he/she comes as a guest, so it is the responsibility of the host country to provide them with every possible facility and comfort because the guest is making expenses for the travel and services. In order to facilitate and promote tourism, to attract the tourist attention, one needs to focus on infrastructure of the country and of all tourist places.

For example, once, foreign tourists' attention is turned on a country, it develops economically by acquiring the foreign exchange which keeps a proper balance of trade. It also raises the standard of living of local people, by giving employment opportunities to unemployed people which helps in their better survival and living. According to a report of 2014, tourism worldwide contributed 3.3% to the world's GDP which gave rise to the world GDP by 10%. Moreover, due to the reduction in the transport costs, traveling became easier hence the GDP is expected to increase more and more in the coming years. The indirect and direct economic impacts of tourism affect the national as well as international borders. For instance, the direct impact of tourism is felt on the sectors like entertainment, food, beverages services, retail opportunities.

On the other hand, the direct impact of tourism occurs inside a country and are fueled by residents and non-residents for business as well as for leisure purposes whereas the indirect economic impact of tourism is apparent in the investments which deal with tourism which cater the interest of private and government sectors. It therefore ensures welfare to the tourist and local stakeholders. It is related to sale, purchase and investments during the period when tourism season is high. The flow of tourist dollars within a community, is another example of indirect impact on the community. Hence, the industry of tourism has affected the economy in destinations of local area, mainly in the less economically developed countries or the developing countries. The expenses made in gaining tourist attention, returns in a form of the income which is generated. International trading capacity expands because of the foreign currencies.

Toll taxes are imposed by the government of the host country. They also earn revenues from taxes on income from tourism jobs, business and entry fee at different tourist sites, heritages and monuments. The materials and items needed by the tourists are usually shipped from other countries. The national and international tourism creates various job opportunities to the informal sectors such as street vendors, rickshaw pullers, tea stalls, Maggie corners, local shops etc. Therefore, the money spent on tourist destinations has improved the lifestyle of local people. This generates the local economy. Therefore, the industry of tourism affects the economy of a country to a large extent.

1.7. POTENTIAL GROWTH IN THE TOURISM INDUSTRY

Tourism consists of tangible and intangible products and services which brings an extreme experience for the tourists. Tourism has evolved from traditional and religious to various other new forms. Over the past twenty years the number of tourists has enlarged and increased to a great extent and an immense growth is expected in the coming years among all the market sectors. It therefore enhances the economy by having an enormous potential to grow and generate revenues. The advancement in technologies has benefited the tour operators and tourists. Moreover, online transactions, booking and navigation has made life easy for people, it has facilitated online travel and accommodations. There are various apps on mobile phones which assists the travelers like a compass app, weather apps, world time apps, language translation apps, google maps etc.

Using these apps one can explore and visit different areas and locations. On the other hand, there are certain apps that let people order food, search for transportation, find inexpensive flights. These apps can maintain the enthusiasm in tourists before, after and in between the tour as it makes their traveling cheap, trouble free and less tiring. Moreover, in recent times, tour operators invest and make their own app. Hence, they connect to people online who want to travel to different locations, they send pdf documents, share details about the trip facilities, budgets and everything else.

They ensure safety and security, find customers, they also give insurance documents to insurance service clients. They make use of a few apps which work as a bridge between them and the tourists, which provide the tourists with all the important information regarding the tour. Some of these apps are Trafalgar app, Passport to Tour app, mTrip app.

Figure 1.9. Potential growth in tourism industry.

Source: Image by unsplash.com.

On the other hand, technology is helping tour operators to create a link between their customers and clients very rapidly. Tour operators are also benefited by this as they promote their business and products. Therefore, they get a chance to know their customers better and provide them an easy way of approach so that they get accessibility of goods and services. Moreover, tourism is getting within everyone's reach and people are enjoying it because of the new technologies. With different apps and online services, it is being made accessible for every sector of society and connecting people from various parts of the world. Nowadays, people are traveling and there are a lot of social media platforms that give them a chance to do travel blogging.

This is increasing day by day and because of this the vogue in people to travel is growing. This has become a part of the trend. Therefore, the availability of options and the affordable prices of every service, the cheap transportation costs have helped tourists to acquire their every dream and wish. Today's generation takes a great deal of interest in technology as well as tourism. They are the best combination for keeping things going in a smooth way. These factors affect the way people travel; it also affects their choice of destinations for vacation.

According to a report given by Google Travel Study, 74% people plan their trip on the internet and 13% seek help from travel agencies to prepare for their tours. Younger generations enhance both the factors as they are zealous about travel and technology both. This is the reason why social media, blogs, apps are gaining rapid popularity. Moreover, people are adapting this trend way too fast. Hence, enthusiastic youngsters are always fond of traveling, they keep on planning trips with their friends, or college trips or education tours. They want to have fun as much as possible and want to live life to the fullest. Therefore, they plan adventure trips, trekking, paragliding and many other adventure sports. Whereas there are some of them who travel to attain peace and connect with nature and to detox themselves. One way or the other the tourism industry is developing at a fastest rate and the indulgence of technology is like an icing on the cake.

1.8. TYPES OF TOURISM

Tourism is a very diversified and complex concept. There are various reasons due to which people travel which eventually leads to the development in the industry of tourism. Therefore, the purpose of the travel of every visitor is different. It can be for personal reasons or for professional purposes. Basically, there are two types of tourism which prevails everywhere. Moreover, they are- Domestic and International. Inbound and outbound tourism is a subcategory of International tourism.

1.8.1. Domestic Tourism

The tourism activity which goes on inside the borders and not across is known as domestic tourism. Domestic tourism is done on a national level, and it is frequent because it does not require any tiring formalities, tasks or a lot of documents. Moreover, traveling in the same country is easier as there is no involvement of visa or passport or regular health checkups or the foreign exchange. Therefore, there are no issues regarding the interaction as there are no barriers of language problems or currency exchange. It is safe and easy to travel in a country as compared to moving anywhere overseas. Moreover, flights work timely, it only needs a ticket to somewhere and people will reach their destination without indulging themselves in formalities.

Domestic tourism helps people know about their country. It records the activities of visitors within their country of residence and outside of their home. They explore their country, cultures, traditions, languages, rituals. It creates a socio-cultural dimension. Therefore, it is marked from ancient times, when people traveled here and there in search of food and shelter. Moreover, the capability of exploring in people never vanished. It has been the very first form of tourism which has been practiced. Hence, pilgrimage plays an important role, as there are numerous devotees and followers who go on pilgrimage in groups to various places.

On the other hand, the long history of domestic tourism traces its steps when people traveled to visit their friends and relatives and as mentioned the religious pilgrimage being the most significant one (Rogerson and Lisa, 2005). Moreover, mass tourism emerged due to the increased disposable income, the establishment of labor rights which are connected to leisure and vacation. Hence, human activities and interaction happening inside a country enhanced tourism.

1.8.2. International Tourism

When the travel happens across the borders, a trip to a foreign country for various reasons, where a valid passport, visa, health documents, foreign exchange is required then this type of tourism is called International Tourism. The continued increase in geographical spread and variations of tourist products and spots is the main characteristic of International Tourism. This type of tourism is proficient in Europe, and America but new tourist markets are noticed in the developing countries as well. America and Europe are the highest tourist receiving countries. Moreover, according to the UNWTO report, these two regions together attracted 70.7% of the world's total tourist arrivals in 2011. However, by the year 2016, this ratio reduced to 66.9% as the tourist attention directed towards the developing countries such as Asia and Pacific. Later by the passing years, the rate of tourism in Europe decreased more rapidly from 44.8% to 36.7%.

On the other hand, international tourism has contributed to a great extent in the world's economy. It has therefore become a crucial source for foreign exchange and employment in various developing countries. Therefore, in recent years international tourism has been given much importance in the developing nation's schemes and strategies. It has therefore set down on the programs of a lot of recent international conferences on sustainable development. Every country has its own characteristic feature, regarding the stay of tourists, their purpose of visit, and geographical distance. These are the factors due to which the growth of the tourists' arrival varies from country to country. Moreover, the highest average tourist arrival was reported in America and then Asia and Pacific. Hence, over the period of 2011 and 2016, countries like America and the Middle East successfully generated the revenues from the international tourists, whereas in Europe, Asia and Africa during 2011 to 2016, the international tourism receipts per arrival decreased gradually.

These reports give a fair idea that countries should not only focus on the tourist arrival but also work for the betterment of their way of living, they need to implement effective and efficient policies, and provide every possible facility. Therefore, they add more value to the products and services provided to the international visitors. Give them better housing and hotel facilities, food and everything at affordable prices. Moreover, it needs work on infrastructure and local communities which are a little backward yet a tourist spot. However, when all the assistance and facilities will be provided to them, tourists will stay longer at their destination. Countries should not make growth in the economy a focal point of the tourism industry, they need to work for the welfare of foreign visitors.

1.8.3. Inbound Tourism

When a foreigner enters a particular country like people traveling outside their native countries to another country it is referred to as inbound tourism for the country where a person is traveling. For instance, when a tourist from Indian origin travels to China, it will be an inbound tourism for China because foreign tourist is coming to China. It is therefore a significant type of tourism. Moreover, there are a lot of countries that depend on the demand from inbound tourists to expand the development and operations of the tourism economy. Inbound tourism is often seasonal; therefore, the destinations will have extreme or low seasons. These factors rely on weather conditions such as sun or snow or school and public holidays.

On the other hand, the World Trade Organization (WTO) and United States (UN) Nations Statistics Division (1994) stated that, "in order to qualify as a tourist one needs to travel and remain in a place outside of their usual residential environment for not more than one consecutive year for leisure, business or other purposes." The important aspect of inbound tourism is bringing a good amount of money to a country through the process of foreign exchange. This is specifically significant to countries where the currency is weaker than the currency of the tourist's home countries. Countries target their promotion and advertising towards certain nationalities.

The country that produces most outbound tourism is China. There are countries that attract Chinese inbound tourists due to the economic value of this market. The dependency on inbound tourism becomes risky and challenging for destinations. There are destinations like Maldives, Spain, and Greece who depend largely on people traveling from other countries to their destinations for tourism.

1.8.4. Outbound Tourism

Tourists who travel from the country of their origin to another country develop inbound tourism. When a person is traveling to a foreign country it is known as outbound tourism for their country as they are moving outside their country. For instance, if a tourist from India is traveling to Europe, then it is outbound tourism for India and inbound tourism for Europe. In other words, the act of traveling out of a native country or home country for the purpose of tourism. It excludes the process of purchasing goods and services before and after the trip within the country that generates tourism.

As stated by the (WTO) World Trade Organization and United States (UN) Nations Statistics Division (1994), that "in order to qualify as a tourist one must travel and remain in a place outside of their usual residential environment for not more than one consecutive year for leisure, business and other purposes." Moreover, outbound tourism is important to many countries as it has a lot of positive effects on the economy and it is so much more than the tourism industry. It generates the economy of various countries by fueling many sectors such as retail, healthcare and education. Many countries are largely dependent on outbound tourism.

1.9. CONCLUSION

Tourism is a rapidly growing industry, yet it is a complex process. A country's economy is largely dependent on the tourism industry. The reason why tourism is growing so fast is because there is a substantial demand from people for tourism. They want to explore, travel to new places in their vacations. Therefore, they spend a large part of their income while traveling. If not for the purpose of traveling, they visit their relatives who live in different cities or sometimes countries. Moreover, the importance of tourism is often instilled in people by advertisement on televisions, or online, mainly on social media and booking sites. Hence, along with such demands there is also another factor called the supply. The tourist facilities are supplied greatly around the world.

Countries are developing their infrastructure to attract the attention of tourists and they have also succeeded to a great extent. They are trying to make the place appropriate for tourism by maintaining restaurants, hotels, markets etc. The policies implemented by the government for promotion, encourage tourism. Also, because of international security tourism offers more destinations worldwide. Moreover, the influence of social media has largely affected the development of tourism as it has taken the place of traditional, time taking methods. It is concerned to individuals, as they are making their own website and blogs and encouraging trips. Hence, they are taking media and promotions in their own hands instead of leaving it to the government and agencies.

Nevertheless, due to the wide use of social media, social networking sites and interaction between people have increased immensely which has developed and encouraged tourism. Another factor that affects tourism is the diaspora growth. Therefore, people living in foreign countries sometimes come back in search of their homeland and roots, to feel the sense of connection with their country and soil. They search for the countries where they were born, and this quest leads them towards their country of origin which eventually results in escalating the overall tourism.

With respect to the rate at which tourism is flourishing and growing, it has a very bright future as an industry especially by (2030). Since the ancient times to recent years, it has developed. Although there are certain negative impacts on the environment and climate which may deteriorate the development. Yet it will not decline in the coming years as steps are being taken for the sustainable tourism industry. Therefore, the forecast done by UNWTO that international tourism will continue to expand at the average annual rate of 4%. Moreover, e-commerce has helped to trade tourist products on the internet. From local economy to the national economy, the tourism industry has enlarged and acted as a bridge for revenues in the country. Further, sustainable tourism has increased the chance of the growth of the tourism industry, by working on environmental, social and cultural impacts on tourism. Hence, this industry needs to be preserved by tackling all the negative impacts it bestows as people look at tourism as an escape, relaxation pursuit of recreation and pleasure away from their home. It is the product of modern social arrangements.

REFERENCES

1. Asadzadeh, A., & Mousavi, M., (2011). *The Role of Tourism on the Environment and Its Governing Law.* [online] Ejbio.imedpub.com. Available at: https://ejbio.imedpub.com/the-role-of-tourism-on-the-environment-and-its-governinglaw.php?aid=19002 (accessed 8 August 2021).
2. Choo, H., (2015). Domestic tourism. *Encyclopedia of Tourism*, [online] pp. 1–2. Available at: https://link.springer.com/referenceworkentry/10.1007%2F978–3-319–01669–6_255–1 (accessed 8 August 2021).
3. Creaco, S., & Querini, G., (2003). *The Role of Tourism in Sustainable Economic Development.* [online] Ideas.repec.org. Available at: https://ideas.repec.org/p/wiw/wiwrsa/ersa03p84.html (accessed 8 August 2021).
4. Dr Hayley Stainton, (2020). Inbound tourism explained | *Understanding the Basics – Tourism Teacher.* [online] Tourism Teacher. Available at: https://tourismteacher.com/inbound-tourism/ (accessed 8 August 2021).
5. Dr Hayley Stainton, (2020). Outbound tourism | *Understanding the Basics – Tourism Teacher*. [online] Tourism Teacher. Available at: https://tourismteacher.com/outbound-tourism/ (accessed 8 August 2021).
6. Iwona Florek, (2012). *Sustainable Tourism Development*. [online] Available at: https://www.researchgate.net/publication/236108688_Sustainable_Tourism_Development (accessed 8 August 2021).
7. Matt Burdett, (2018). The future of tourism – GeographyCaseStudy.Com. [online] GeographyCaseStudy.Com. Available at: https://www.geographycasestudy.com/the-future-of-tourism/ (accessed 8 August 2021).
8. Prince Kumar, (2021). Tourism and Its Importance: Introduction to Tourism, Hospitality; Hotel Industry – *hmhub*. [online]. Available at: https://hmhub.me/tourism-importance/ (accessed 8 August 2021).
9. Sustainabledevelopment.un.org. n.d. Sustainable tourism: Sustainable Development Knowledge Platform. [online] Available at: https://sustainabledevelopment.un.org/topics/sustainabletourism (accessed 8 August 2021).

10. Tutorials point, (2021). Tourism Management – *Trends and Future – Tutorialspoint*. [online] Tutorialspoint.com. Available at: https://www.tutorialspoint.com/tourism_management/tourism_management_trends_and_future.htm (accessed 8 August 2021).

Chapter 2

Tourism Demand Forecasting: An Overview

CONTENTS

This chapter explains the basic overview of the concept or ideology behind tourism demand forecasting. This chapter also explains an overall view on forecasting international tourism. It therefore provides highlights on the various types of techniques and methodologies which aids in forecasting the tourism demand. This chapter also emphasizes on different types of demand forecasting such as active demand forecasting, passive demand forecasting, long haul projections, external macro forecasting, and internal business forecasting. It further sheds light on the various sorts of strategies that have been used in demand forecasting which includes, trend projection, market research, salesforce composite, and so on. It thus addresses the various kinds of factors that are influencing tourism. This chapter also explains several social as well as safety factors with respect to the demand forecasting in tourism.

2.1. INTRODUCTION

The principal objective of the travel industry request estimating is to help destinations and the travel industry organizations to keep up ceaseless supplies of the tourism items and administrations to fulfill the expanding interest for global travel experiences. Therefore given the intricacy of tourists' conduct in choosing destinations and the travel industry items / administrations, extensive endeavors have been made to comprehend the components that decide the interest for the travel industry. These incorporate destination/product-specific variables, like the costs of the travel industry items and services at destinations, travel costs from cause nations/districts to destinations and expenditure use by destinations. They likewise incorporate source market-and tourist explicit variables, for example, the social and monetary states of the source markets and sightseers' salaries.

The capacity to anticipate such factors is significant in precisely forecasting future interest for the travel industry at the destination and product levels. An enormous number of the travel industry forecasting studies have in this manner focused on the details of the travel industry demand models by taking all conceivable affecting elements into thought when building demand models. Moreover, an additional benefit of this line of examination is that the discoveries are valuable for strategy and chiefs in planning fitting arrangements and techniques to energize and oversee future interest vacillations.

Figure 2.1. Tourism demand forecasting.

Nevertheless, time arrangement and artificial intelligence (AI) models have additionally shown up every now and again in the tourism forecasting literature. Their essential goals are to improve estimating exactness and limit the expense of looking for far reaching causal models. They can likewise be utilized as benchmarking models to contrast determining execution and causal models. Therefore, judgmental forecasts have been utilized in the travel industry by estimating when the notable information for certain travel industry items/administrations are missing and when it is important to change the factual or econometric forecasts of the travel industry forecasts.

The primary tourism demand study was published in 1958. From that point forward, more than 670 studies on tourism demand modeling and forecasting have appeared in English plan forecast academic journals. However, during the 1960s and 1970s, the static relapse approach was predominant, and the travel industry demand studies were principally worried about researching the determinants of demand. Moreover, additional forecasting models were applied during the 1980s as specialists considered the time-arrangement construction of the travel industry demand information. Around there, a few researchers utilized static relapse models, and they considered improving time-arrangement models (e.g., credulous, autoregressive and dramatic smoothing models and pattern investigation).

On the other hand, during the 1990s, the quantity of utilizations of time-arrangement models proceeded to increment, and models dependent on unique details likewise filled in notoriety (e.g., the Box–Jenkins's strategy with exogenous factors and the autoregressive dispersed slack model (ADSM)). Moreover, framework based econometric models (for example, the vector autoregressive (VAR) model, the practically ideal interest framework (AIDS) and AI-based models) additionally arose. This pattern proceeded into the 2000s, with impressive new advancements in present day econometric models, AI-based models and joined and half-breed strategies.

Nevertheless, demand forecasting is the way toward assessing the amount of an item or administration that shoppers will buy. Demand forecasting essentially includes techniques including both casual strategies as suppositions, and quantitative strategies, like the utilization of chronicled deals information or current information from test markets. It might, therefore, be utilized in settling on estimating choices, in surveying future limit prerequisites, or in settling on choices on whether to enter another market. The estimate of the travel industry of uncommon significance since it is a marker of future interest, along these lines giving fundamental data to resulting arranging and strategy making. However, the fundamental determining strategies which are utilized in the travel industry and diversion fields are time-arrangement techniques, numerous relapse strategies, multivariate strategies and subjective estimating strategies. Studies which look at the exactness of forecasts of the travel industry and amusement request created by the different methods are then analyzed.

The travel industry figures might be produced by either quantitative or subjective methodologies. Quantitative methodologies create and utilize numerical models, speculations and theories relating to regular marvels. In contrast to quantitative methodologies, subjective methodologies include top to bottom comprehension of human conduct. Practically all forecasting includes foreseeing the travel industry interest sooner or later. Therefore, as an extravagance decent, the interest for the travel industry will in general be very flexible while the pay flexibility of various the travel industry items can contrast significantly, as some diversion merchandise may really show declining utilization with expanding pay.

2.2. AN OVERALL VIEW ON FORECASTING INTERNATIONAL TOURISM

Forecasting is a subject that entrances numerous individuals who are keen on the financial aspects of the travel industry. Evaluations of future interest at the objective level are vital in overseeing and arranging the travel industry advancement and the essential speculation. Nonetheless, forecasting in the travel industry area is certainly not a simple work.

The travel industry business analysts have created forecasting models to foresee travel interest for each major worldwide market and specific voyager portions. More significant, these models have set up demonstrated histories for exactness. The explanation is clear and the models are solidly established in the financial essentials of source showcases alongside the changing elements of the travel industry and voyager inclinations.

During the time spent arranging, forecasting happens in sync "point by point examination and investigation" and the phase of "engineered" is communicated when the assessed sum that is proposed to reach through the arrangement.

Figure 2.2. An overall view on forecasting the international tourism.

The travel industry forecasting approach incorporates all strategies utilized in the determining interaction. This system should therefore be logical on the grounds that it can just acquire helpful outcomes. Regardless of its significant job, the action of forecasting the travel industry hates consistent acknowledgment of its significance in the arranging interaction and in figuring macroeconomic approaches. This is on the grounds that first, most transient conjectures are made and consequently cannot be a help for key arranging and forecasting philosophy then again, the travel industry is less evolved than other monetary areas.

On the other hand, under the name "The travel industry 2020," the WTO set up a drawn-out estimate of the travel industry interest in making an evaluation of its improvement until (2020). It is a quantitative Forecast. Therefore, covering 25 years (requiring the base year 1995) and disintegrates the stream in the 6 territories overall travel objective: Europe, Americas, East Asia and Pacific, Africa, Middle East and South Asia. Moreover, despite the way that travel industry improvement lately has been sporadic, UNWTO keeps up this drawn-out vision, considering involvement that times of solid development (1995, 1996, and 2000) are trailed by periods when development is slower, however the overall pattern remains practically unaltered. Hence, the WTO forecast for 2020 gives various appearances 2.56 billion, of which 2.2 billion will be intraregional and 0.4 billion travel significant distances. However, breaking down the absolute number of appearances by locale shows that the main territories are Europe – 717 million sightseers, East Asia and Pacific – 397 million tourists, Americas –

282 million tourists, trailed by Africa, Middle East and South Asia. Hence, development paces of around 5% each year, more than the world normal of 4.1% will be recorded in Asia, Middle East and Africa. Europe and America are as of now profoundly created locales as far as the movement to encounter development rates underneath the worldwide normal.

On the other hand, Europe will have the most elevated level of appearances, regardless of whether it will tumble from 60% in 1995 to around 46% in 2020. Global significant distance travel will develop at a normal yearly pace of 5.4% and intraregional streams in a normal of 3.8%. In this manner, if intraregional travel in 1995 represented 82% of the absolute worldwide significant distance and 12%, in 2020 the rates will be 76% to 24%. Moreover, when creating a virtual tourism strategy, an unmistakable qualification ought to be made between the virtual market of items deals and the virtual market of administration deals. Inside the regular virtual business sectors, one may work through selling items (like books, blossoms, metals, officialdom and so forth) and with two sorts of members: providers and clients.

The travel industry items accessible inside the virtual business are, for the most part, items expected for current use. Therefore, next to the providers and the clients inside the virtual the travel industry advertises that sell such sort of items, one may likewise accentuate the presence of certain social orders that guarantee support, for example, the banks or the organizations liable for conveyance.

With innovation, correspondence and operational expenses are diminished and therefore, adaptability, intelligence, proficiency, efficiency and intensity are improved. information innovation has been liable for driving the shift from product orientated tourism organizations to more adaptable and responsive market-orientated tourism activity, where achievement relies upon detecting and reacting to quickly changing client needs. We can thus recognize three essential types of networks utilized in the travel industry:

- Internet works with the intuitiveness of the endeavor and people with the whole scope of the outside world through interactive media representations.
- Intranet is shut, gotten or 'fire walled' networks inside associations, which tackle the requirements of inward business clients, by utilizing a solitary controlled, easy to use interface to exhibit all organization information.

- Extranet uses a similar rule with outside PC organizations to improve the intelligence and straightforwardness among associations and confided in accomplices.

This advancement worked with (Internet and World Wide Web) an exceptional chance for circulation of media data and intuitiveness among directors and purchasers. Nonetheless, the data accessible on the Internet has presented new practices is expected that the virtual home and venture will arise. It additionally engages the shopper through the arrangement of customized items that meet their individual requirements, so overcoming any barrier between the purchaser and objective in an adaptable and intuitive manner. Moreover, the Internet can likewise fortify the showcasing and correspondence elements of distant, fringe and separate objections, just as little and medium-sized tourism ventures, by empowering direct correspondence with forthcoming clients.

Henceforth, the Internet gives extraordinary and moderate freedoms to the worldwide portrayal and showcasing for both huge and little tourism providers. Pragmatic it gives special freedoms to interactive media offer which addresses the new purchaser of the travel industry and settled the follows issues:

- Increase the security of transmissions.
- Ensure validity and responsibility of data.
- Secure protected innovation and copyright issues.
- Enhance data transfer capacity and decrease speed restrictions.
- Reduce client disarray and disappointment.
- Provide enough prepared trained professionals.
- develop equivalent access for more modest and bigger accomplices.
- establish estimating structures for the appropriation of data and reservations.
- enhance the normalization of data and reservation methodology.

While the development in the quantity of Americans arranging trips online has remained moderately steady, there has been solid development in online travel booking. To be sure, four of every 10 (41%) online voyagers are reserving or reserving travel spots on the web. This may therefore incorporate booking an aircraft ticket, lodging, rental vehicle or bundle visit on the web. Virtually those who reserve travel spots online say they reserved a spot for joy, get-away, or individual purposes. Around three out of 10 say they reserved a spot for business or show purposes.

> The actual meaning of sustainable development remains hidden sometimes, especially when it comes to tourism. Although the concept of sustainable development is a global concept yet there is a failure to acquire and work upon the real meaning of this term.

As the Internet progressively turns into the standard instrument for movement arranging, travel services can utilize the Internet openings for advancement, yet in addition for new purposes, like gathering and exploring data and accessing new colleagues, providers, and administrations. Simultaneously, the Internet is essentially utilized as a wellspring of data and item and value correlation instruments, particularly in Europe. Not all people visiting travel sites additionally book on the web, while others, from the earliest starting point, incline toward the individual contact either vis-à-vis, or by means of telephone or email. Moreover, the utilization of the Internet by explorers to plan and book their outings keeps on developing at a fast rate. In the United States, as indicated by the tourism Association, in excess of 75 million online voyagers utilized the Internet in 2009 to get data on objections or to check costs and timetables.

From 2006 to June 2009, web-based booking showed a wonderful twofold digit development for four back-to-back years, with a terrific 58% development in 2006, trailed by a 25% development in 2007. While the development of the online voyager's market has eased back, the quantity of online explorers who really utilize the Internet to design trips has remained moderately steady. A larger part (67%) of online voyagers say they counsel the Internet to get data on objections or to check costs or timetables. For Romania, the World Tourism and Travel Council has made a figure for the time frame 2007–2016 in the "travel industry advancement technique" that gives normal paces of development to the travel industry:

- 7.4% commitment to GDP, better than expected in the EU is 2.4% and the world which is 3.2%.
- 2.7% regarding occupations in tourism, top EU normal is 2.5 % and below normal which is 2.8%.

- 7.9% for tourism and travel interest and higher EU normal of 3.5% and the normal worldwide is 4.2%.
- 8.5% tourism accepting higher EU normal of 4.3% and world normal of 4.9%.
- 6.2% on the capital venture, higher EU normal of 4.2% and 4.6% worldwide normal.

As per these information, Romania positions 174 nations analyzed in 4 of the development paces of long-haul interest. To these will help improve Romania's picture considering NATO and EU promotion in 2007. In 2016 states that travel industry and travel industry will accomplish USD 6.2 billion, arriving at a commitment of 2.4% of GDP. Moreover, several individuals utilized in tourism in 2016 will be 315 000 individuals, for example 3.83% of all out business. Additionally, the Tourism Forecasting Committee (TFC) was set up after the Tourism White Paper's drive to frame Tourism Australia (TA). Like its archetype the Tourism Forecasting Council, the TFC stays a free body accused of giving present and potential tourism financial backers, industry and government with agreement figures of action across global homegrown and outbound tourism areas. The assets to run the TFC are given by Tourism Research Australia (TRA).

By agreeing with Tourism White Paper's following quite a while of solid development, world financial development tumbled to a beneath pattern pace of 3.2% in 2008, including a sharp fall in activity in the second 50% of the year.

In the last arrival of Forecast (in December 2008), the International Monetary Fund (IMF) estimate for world financial development in 2009 had tumbled from pattern development (or around 4.0%), to about a large portion of that rate (2.2%). From that point forward, the world macroeconomic viewpoint for 2009 has disintegrated, and thus, the pace of decay has been emotional. Moreover, in its April 2009 proclamation, the IMF figures a compression on the planet economy of 2.3% in 2009, with a re-visitation of humble development of 2.9% in 2010. Further, joblessness is rising pointedly in most western economies and for its conjecture to stay high for certain years. Joblessness levels will be a central point impacting the circumstance and strength of worldwide recuperation.

Nevertheless, conjectures for tourism movement in 2009 for Australia's other inbound business sectors have been downsized significantly, including New Zealand, Japan, the United States, the United Kingdom, Germany, Singapore, Malaysia, South Korea and Hong Kong. The biggest descending corrections to the 2009 estimate are for Japan (presently conjecture to decay by 6.6% in 2009) and Singapore (–7.0%). Thus, International tourism will keep on confronting considerable difficulties in 2009 and 2010, as just a lethargic recuperation is normal from the current world financial downturn. The travel industry is straightforwardly impacted by financial downturns as optional spending is diminished while investment funds rates are expanded in accordance with low purchaser (and business) certainty.

As detailed by the United Nations World Tourism Organization (UNWTO Barometer, January 2009), worldwide tourism expanded emphatically in the principal half of a year ago (up 5%), however fell by 1% in the second 50% of 2008 as most the world's driving economies contracted due to the worldwide monetary emergency. Most of the lessening in appearances in the second 50% of the year was accounted for in Asia and Europe (both down 3%), while development proceeded in the Americas, (up 1%), Africa (up 4%) and the Middle East (up 5%). For a similar period, global guest appearances to Australia diminished by 2.5%.

2.3. FORECASTING TECHNIQUES IN TOURISM DEMAND

This rundown is centered around showing the forecasting strategies used to decide the probably request in tourism and contends that given the significance of tourism area to the economy of any tourist country. Therefore, precise figures of tourist appearances are of significance for arranging by both the private and public areas. First, one should address the inquiry what tourism is.

Clearly the travel industry is not just one organization. It joins a huge number of items and administrations. An organization defines objectives and utilizations its creation, showcasing and administrative assets to accomplish them through its administration interaction. Therefore, it ought to be referenced that a fall popular can achieve diminishes in expectations for everyday comforts following the ascent in joblessness, while expanded interest can prompt higher business, pay, yield and expansion too may undermine natural quality and maintainability. In addition, tourism firms are stood up to by changing income and benefits and governments experience changing duty income and use. In this way, tourism request impact can be seen in all areas of economy – families and people, public area and private organizations. For instance, choices on tourist uses, tourism markets design and dynamic nature between them, cross-country linkages between tourism firms, the commitment of ecological assets and their significance to approaches for reasonable tourism have not yet been completely researched and need further financial investigation.

The chapter is pointing on showing the current forecasting strategies, their positive and negative highlights for better understanding, the significance of interest estimating in tourism, and the need of utilizing these or those techniques for getting the most exact and exact outcomes. Clearly, one of the more intricate parts of tourism is the travel industry interest. Therefore, when in doubt it is characterized and estimated in an assortment of ways and at a scope of scales. For the most part, there are financial, mental and social mental techniques utilized in forecasting. For instance, choice to buy occasions are frequently made with loved ones so shopper request hypothesis dependent on singular dynamic should assess people's and gatherings' social settings.

Just as the examination of movement examples and modes has been overwhelmed by geological insightful systems, while the investigation of interest outside financial aspects will in general be supported by mental or social mental strategies. Therefore, the numerous investigations of tourism interest in various nations and timeframes are investigated by Archer, Johnson and Ashworth, Sheldon and Sinclair while Witt and Martin analyzed elective ways to deal with the travel industry request forecasting. Moreover, the meaning of the travel industry request gives a solid case to better comprehension of the dynamic cycle nature among sightseers. Hence, if there should be an occurrence of utilizing an improper hypothetical system in observational investigations of interest can bring about off base detail to forecast tourism interest and one-sided proportions of the responsiveness of interest to changes in its determinants.

It ought to be referenced that 'observational examinations help to clarify and comprehend the level and example of tourism interest and its affectability to changes in the factors it is dependent on. For instance, it helps in noticing pay in starting point zones, trade rates between various objections and inceptions just as relative paces of swelling. This sort of data is of significance to public area strategy making and the private area.' In any case, just if there should be an occurrence of suitable hypothetical detail of the considering model utilized the appraisals can be exact and exact. Thus, unequivocal thought of the shopper dynamic supporting experimental models is of significance in introducing the gave forecasts are neither deceiving nor incorrect in their arrangement suggestions. Consequently, there are two methodologies used to show tourism interest.

The initial one is the single condition model and the second is the arrangement of the condition model. 'The first single condition model has been utilized in investigations of tourism interest for various nations and time spans and expresses that request is a component of various deciding factors. ' (Sinclair, 1997). This method allows the estimation of the interest affectability to changes in these factors. In opposition to the principal approach, the arrangement of conditions model requires the synchronous assessment of a travel industry request conditions range for the nations or kinds of tourism consumption considered. Therefore, arrangement of conditions procedure attempts to clarify the affectability of the spending portions of tourism interest across a scope of sources and objections (or tourism types) to changes in the basic determinants.

Nevertheless, there exists one seriously forecasting method which is later and can measure up to the single condition approach. It is the Almost Ideal Demand System (AIDS) (Maria De Mello, 1999). This model is hypothetically better compared to the referenced above and offers a scope of valuable data concerning the affectability of tourism interest to changes in relative costs and in sightseers' use spending plan. This methodology was utilized in analyzing the UK interest for tourism in its topographical neighbor-nations as France, Spain and Portugal. Moreover, the aftereffect of such examination showed that 'the UK interest for tourism in Spain expanded more than proportionately regarding an ascent in the UK use spending plan for tourism in three nations, the interest in tourism in France expanded not exactly proportionately and the interest in tourism in Portugal stayed stable. The affectability of the UK interest for tourism in Spain to changes in powerful costs in Spain is expanding and surpasses the relating upsides of the sensitivities of the interest in tourism in France and Portugal to changes in French and Portuguese costs, individually.' Conversely, the UK interest for tourism in Spain is inhumane as for changes in costs in its more modest Portuguese neighbor.

The UK interest for Portugal is delicate to changes in costs in Spain, albeit the level of affectability gives off an impression of being diminishing over the long haul, and the interest in France (Portugal) is inhumane concerning an adjustment of costs in Portugal. So, clearly dependability of interest even with rising costs might be seen as signs of accomplishment, and opposite results mean a potential case for reconsidering strategy toward tourism interest. Researchers have utilized an assortment of other forecasting strategies during the previous a long time for the tourist industry. Among them are quantitative determining techniques. They might be characterized into two classes: causal strategies (relapse and primary models) and time arrangement techniques (fundamental, middle of the road, and progressed explorative techniques).

For additional clarification we should specify that causal techniques address procedures for distinguishing connections among autonomous and subordinate factors and endeavor to fuse the associations of different factors. Notwithstanding, there is sure trouble of applying the causal techniques. It is distinguishing the free factors that influence the estimate factors. Thus, the precision and dependability of definite estimate yields made under causal techniques rely upon the nature of different factors. The second gathering of strategies, time arrangement quantitative techniques, offers numerous benefits. It is brought up that 'the utilization at time t of accessible perceptions from a period arrangement to conjecture its worth at some future time t+1 can give a premise to:

- financial and business arranging,
- creation arranging,
- stock and creation control, and
- control and advancement of mechanical processes.

Time arrangement strategies offer methods and ideas working with determination, assessment and assessment. They gain more exact forecasting results than those yielded by causal quantitative strategies. It therefore ought to be referenced as an illustration that forecasting is muddled by the solid irregularity of most tourism arrangements.

On the other hand, it is brought up that to consider irregularity to be a type of information pollution is one of the commonplace ways to deal with the investigation of macroeconomic time arrangement. This was the methodology frequently utilized in many registration and factual offices. Therefore, an account of the tourism analysis seasonality is indispensable to the interaction and is of high significance for the circumstance of the issuance of strategy measures as well as examining the since a long time ago run pattern.

'As huge highlights of quantitative tourism forecasting (including the mathematical examination of verifiable information) we see that while it is especially valuable for existing tourism components, it is restricted in its application to new ones where no past information exists' (Smith, 1996). This procedure was utilized in determining potential UK interest for space tourism. Therefore, univariate forecasting strategies might be utilized to forecast appearances. This restricted system comparative with underlying models permitting strategy producers to decide how changes specifically factors can assist with improving the business. Hence, the flimsy spot of the strategy is that the models have no informative factors, so it is hard to decipher the individual parts. Hence, the forecasting record of numerous univariate models has impressive forecasting precision. Lim and McAleer utilized univariate methods to forecast quarterly tourist appearances to Australia and to decide their estimating exactness utilizing an assortment of occasional channels. Kulendran and King additionally utilized an assortment of models to rank determining execution of different tourist appearance arrangements utilizing occasional unit root testing.

2.3.1. Ends and Recommendations

It ought to be referenced that determining procedures and estimating itself have some flimsy spots. Right off the bat, current determination is for the most part the area of strategy producers. Therefore, it is advantageous for three gatherings: public area tourism associations as it legitimizes spending distributions; chiefs of public and private area tourism projects as they may energize financial backers, and the actual forecasters. However, there are no real advantages from determining for tourism administrators and providers because the outcomes are not significant and random to the genuine necessities of most of the tourism organizations. The issue with the outcomes might be outlined by such a model. The BTR's "Australian Tourism Forecasts" report delivered in April 1990 forecasts 4. 85 million guests constantly 2000.

Nevertheless, BTR's most recent forecast for 2000 is 4. 824 million guests. What is more, just a month ago The Australian paper announced the aftereffects of "a regarded private area forecaster" who has forecast 5. 33 million before the finish of the decade. So, an individual sees numbers continue to change, and this is the proof that occasionally the forecasting results become not significant. Summing up all the referenced above we may say that there is a wide scope of strategies utilized for determining requests in tourism. In this paper, we referenced just some of them which to our brain merit consideration and might be utilized in forecasting the interest.

As you may see examination of tourism demand includes explicit issues since it has some uncommon nature credited to the intricacy of the inspirational construction concerning dynamic cycle just as the restricted accessibility of the important information for econometric displaying.

Moreover, quantitative methodology for tourism demand needs the system of a formal numerical model giving assessments of affectability to changes in the factors the interest relies upon. Econometric displaying offers a decent reason for precise forecasting which is vital to the public area making interests in the business. The single condition model regularly overlooks the unique idea of tourism interest, dismissing the likelihood that the affectability of tourism interest to its determinants may vary between timeframes. The elective model is the Almost Ideal Demand System. It is initially evolved by Deaton and Muellbauer. This model not just allows the assessment of the total arrangement of applicable versatilities, yet additionally takes into consideration formal trial of the legitimacy of the suspicions about buyer conduct inside the example set of perceptions.

The AIDS permits to test suspicions and forecast boundaries in a manner that is absurd with other elective models. So, for the time being, we may say that there are no totally terrible or great procedures utilized for estimating tourism interest. In any case, there are ideal models for getting more precise outcomes. It is better utilizing models dependent on old hypothetical information yet with recent fads ready to cover every one of the important perspectives in forecasting tourism interest.

2.4. TYPES OF DEMAND FORECASTING

Nobody can anticipate the future with total assurance. Nonetheless, there are a few interest forecasting strategies that help you make an informed theory. Utilizing an estimating model will help you settle on better business choices. Therefore, there are a few unique approaches to do demand forecasting. Your figure may contrast dependent on the forecasting model you use. The best practice is to do different demand conjectures. This will give you a more balanced image of your future deals. Utilizing more than one estimating model can likewise feature contrasts in forecasts. Those distinctions can highlight a requirement for more examination or better information inputs.

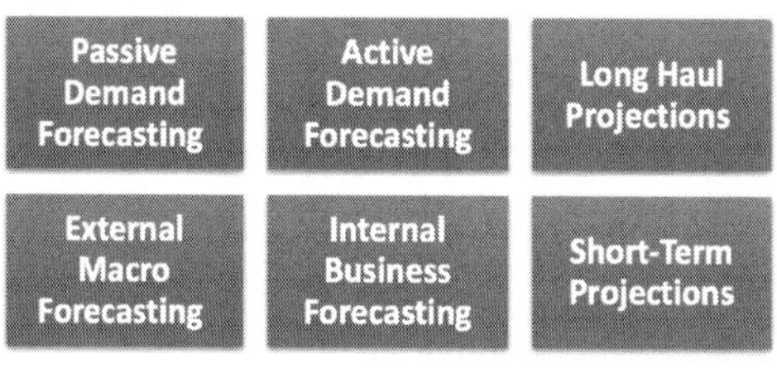

Figure 2.3. Types of demand forecasting.

2.4.1. Passive Demand Forecasting

Passive demand forecasting is the easiest sort. In this model, you use deal information from the past to forecast what is to come. You should utilize information from a similar season to extend deals later, so you make a valid comparison. This is especially evident if your business has occasional changes. Therefore, the detached estimating model functions admirably if you have strong deal information to expand on. Moreover, this is a decent model for organizations that focus on dependability instead of development. It is a methodology that accepts that the current year's deals will be roughly equivalent to a year ago deals. Hence, uninvolved demand determining is simpler than different kinds since it does not expect you to utilize measurable techniques or study financial patterns.

2.4.2. Active Demand Forecasting

If the business is in a development stage or in case you're simply beginning, Active demand forecasting is a decent decision. A functioning estimating model thinks about your statistical surveying, advertising efforts, and development plans. Dynamic projections will frequently factor in facades. Contemplations can incorporate the monetary viewpoint, development projections for your market area, and extended expense reserve funds from production network efficiencies. New companies that have less verifiable information to attract on should put together their suppositions with respect to outer information.

2.4.3. Short-Term Projections

Momentary demand forecasting takes a gander at the following three to a year. This is helpful for dealing with or without a moment to spare store network. Seeing transient demand permits you to change your projections dependent on continuous deals information. It therefore assists you with reacting rapidly to changes in client demand. Moreover, on the off chance that you run an item arrangement that changes as often as possible, seeing transient demand is significant. For most organizations, notwithstanding, a momentary conjecture is only one piece of a bigger riddle. Hence, an individual will most likely need to watch farther with medium-or long-haul demand forecasting.

2.4.4. Long Haul Projections

Your drawn-out figure will make projections one to four years into what is to come. This forecasting model spotlights on molding your business development direction. While your drawn-out arranging will be founded halfway on deal information and statistical surveying, it is likewise optimistic. Think about a drawn-out demand figure as a guide. Utilizing this forecasting procedure, you can design out your showcasing, capital speculations, and production network tasks. That will assist you with getting ready for future demand. Being prepared for your business development is essential to getting that development going.

2.4.5. External Macro Forecasting

Outer large scale estimating joins patterns in the more extensive economy. This projection takes a gander at what those patterns will mean for your objectives. An outer full scale request figure can likewise provide the guidance for how to meet those objectives. An organization might be more put resources into dependability than extension. In any case, a thought of outside market influences is yet vital for your business projections. Outer large-scale figures can likewise address the accessibility of crude materials and different components that will straightforwardly influence your production network.

2.4.6. Internal Business Forecasting

One of the restricting components for your business development is the inner limit. If you project that client request will twofold, does your venture have the ability to fulfill that need? Inner business request conjectures survey your tasks. The inside business forecasting type will uncover impediments that may moderate your development. It can likewise feature undiscovered spaces of chance inside the association. This estimating model variable in your business financing, cash available, overall revenues, inventory network tasks, and staff.

Inward business request forecasting is a useful instrument for making practical projections. It can likewise guide you to zones where you need to construct limits to meet development objectives.

2.5. STRATEGIES FOR DEMAND FORECASTING

There are various approaches to make figures. Here are five of the top interest forecasting strategies.

Figure 2.4. Strategies for demand forecasting.

2.5.1. Trend Projection

Pattern projection utilizes your previous deals information to project your future deals. It is the least complex and most direct interest forecasting technique. It is essential to change future projections to represent verifiable oddities. For instance, maybe you had an unexpected spike sought after a year ago. Nonetheless, it occurred after your item was included on a famous network show. Therefore, it is probably not going to rehash. Or then again, your eCommerce site got hacked, making your business plunge. Make certain to note uncommon elements in your verifiable information when you utilize the pattern projection technique.

2.5.2. Market Research

Statistical surveying request estimating depends on information from client studies. It requires time and exertion to convey studies and arrange information, yet it is awesome. This technique can give significant experiences you cannot get from inside deals information. Moreover, this exploration can be done on a progressing premise or during an escalated research period. Statistical surveying can give you a superior image of your normal client. The studies can gather segment information that will help you target future promoting endeavors. Hence, statistical surveying is especially useful for youthful organizations that are simply becoming more acquainted with their clients.

2.5.3. Salesforce Composite

The business power composite interest estimating strategy puts your outreach group steering the ship. It utilizes criticism from the business gathering to conjecture client interest. Therefore, your sales reps have the nearest contact with your clients. They hear input and take demands. Subsequently, they are an extraordinary wellspring of information on client wants, item drifts, and what your rivals are doing. Moreover, this technique assembles the business division with your supervisors and chiefs. The gathering meets to build up the figure collectively.

2.5.4. Delphi Technique

The Delphi strategy, or Delphi procedure, use well-qualified conclusions on your market forecast. This strategy requires drawing in external specialists and a gifted facilitator. Therefore, an individual can start by sending a poll to a gathering of interest forecasting specialists. You make an outline of the reactions from the first round and offer it to your board. This cycle is rehashed through progressive rounds. Moreover, the appropriate responses from each round, shared secretly, impact the following arrangement of reactions. The Delphi strategy is finished when the gathering goes to an agreement.

This interest determining technique permits you to draw on the information on individuals with various subject matters. The way that the reactions are anonymized permits every individual to give straightforward answers. Moreover, since there is no face-to-face conversation, you can incorporate specialists from any place on the planet on your board. The cycle is intended to permit the gathering to expand on one another's information and assessments. Hence, the outcome is an educated agreement.

2.5.5. Econometric

The econometric strategy requires some calculating. This strategy consolidates deal information with data on external powers that influence interest. At that point you make a numerical recipe to foresee future client interest. The econometric interest forecasting strategy represents connections between financial components. For instance, an increment in close to home obligation levels may match with an expanded interest for home fix administrations.

2.6. FORECASTING FOR TOURISM AND ITS PRODUCTS

"Forecasting" signifies a future forecast. It is very fundamental to have the option to forecast modern advancements in the arranging interaction. Since an arrangement is worked out, in view of specific presumptions. These suspicions are expected based on determining. Particularly in promoting, to plan an advertising plan exact forecasting is urgent. The conjecture may foresee certain conduct of clients or explicit moves arranged by contenders. One needs to likewise consider the arrangements of government or general financial, lawful or natural improvements which lead to particular patterns.

Figure 2.5. Forecasting for tourism and its product.

For instance, the quantity of individuals making a trip to explicit objections will change as indicated by certain irregularities. If we think about friendly irregularity, we realize that more individuals will go during get-away period. Moreover, if one considers climatic irregularity, we realize that in summer individuals from blistering fields will need to disappear to slope stations to appreciate cooler air. Hence, the visa guideline strategies of the government may influence several global sightseers. The conversion scale will influence the quantity of inbound or outbound tourists. Subsequently to comprehend the conceivable conduct of clients required to consider different factors and make a conjecture which will help in planning reasonably exact plans.

Appropriately in high tourist season the travel industry ought to be prepared to get a bigger number of tourists, and in low season key ought to plan for low deals turnover. Additionally, since the items in the travel industry are different administrations, they have attributes that make it hard to stock the item. In any case, the business needs to set itself up for highs and lows of interest differently. It could therefore be said that forecasts of the travel industry request are fundamental for effective arranging via aircraft, rail lines, hoteliers, visit administrator's tourist transport administrators, food and catering foundations and different businesses associated with or subject to the friendliness and the travel industry markets. This Unit acquaints you with various factors in forecasting, various techniques in forecasting and the significance of estimating in the travel industry.

2.6.1. Forecasting for Tourism

Tourism is a multi-layer industry which other than the objective has three primary areas: Transport, Accommodation and Intermediary administrations. Tourism "Items" are straightforwardly reliant upon the general turnover of tourism which thusly relies upon and its Product number of tourists and the cash they spend. Likewise, however these areas of the business are offering various items, they are intently reliant on one another to maintain the business easily.

For instance, if tourists and visit administrators do the advancement, tourists are attracted to the location, who will utilize the vehicle administrations and afterward convenience and nearby administrations. So, the advancement will profit tourist or visit administrators as well as transport and convenience area. Moreover, in the event that an objective doesn't have great vehicle benefits, the convenience area at the objective may not get adequate business and bad habit a-versa.

Aside from reliance of these principal areas on one another there are various other complex reasons which influence the patterns in tourism. In any case, basically for showcasing forecasting is needed to work out the Product Design and to comprehend the interest for that specific item. Indeed, forecasting is an essential segment in the dynamic cycle for arranging, sorting out or promoting of tourism items and administrations. For instance, if a carrier is beginning its administrations on another area it will need to realize the traffic working on that area and the sort of traffic. At that point only they can design the recurrence and nature of their administrations. Additionally, for an inn, the number and sort of clients will be important to work out the item offers. Subsequently, for tourism the significant variables are number of explorers and sort of voyagers. One should realize that at some random time in future what volume of tourists you expect and to the amount you can cater.

When this general situation is forecasted, at that point the individual associations can work out their own showcasing blend of Product, Price, Promotion and Place and contend with one another for piece of the pie.

Thus, forecasting is expected to design, create and work tourism offices and administrations. Different benefits of determining are that it can help:

- minimize the danger of oversupply,
- in ideal usage of accessible assets,
- maximize tourism interest,
- and direct the convergence of sightseers – as indicated by the conveying limit.

The World Tourism Organization issues tourism estimates at a worldwide level making projections regarding tourist creating areas and location locales. Be that as it may, this activity is totally disregarded in the space of homegrown tourism in India.

2.6.2. Factors Affecting Tourism

To foresee tourism request it is fundamental to comprehend different components which will influence the quantity of explorers and the progression of voyagers. A portion of these components are:

- Politics and legitimate climate,
- Social and security factors,
- Geographical and climatic variables,
- Accessibility and convenience,
- State of economy, and
- New improvements in industry and so forth.

Governmental issues and Legal climate are one significant factor that influences the developments of tourist streams. Political conditions, the kind of government and travel guidelines (like visa, unfamiliar trade and so on) impact the free development of voyagers. For instance, China has as of late just supported inbound tourism; In India additionally, there are sure locales that are not open to sightseers or one requirement a license to enter those districts. Indeed, the overall political state of a country generally decides the inflow of tourists.

2.7. SOCIAL AND SAFETY FACTORS

Social and Safety factors are straightforwardly identified with the mental and actual security of the tourists. These components likewise influence the patterns of tourist streams. For instance, the progression of tourists to the Kashmir valley has gone done whereas Himachal Pradesh has gotten.

Figure 2.6. Social and safety factors.

Certain social exercises pull in tourists. In the event that there is an extraordinary get-together like a celebration, meet or a gathering the tourist business at that area ought to be prepared to get higher tourist inflows.

2.7.1. Geographical and Climatic Conditions

Geographical and Climatic conditions choose the irregularity of streams, for example, tourist stream from spaces of warm fields will go to slope stations in summer or sightseers from colder nations will get away to radiant seashore resorts during winter and so forth by mentioning appropriate observable facts the forecasting of tourist appearances can be made to embrace arranging of extra tourist inflows.

2.7.2. The Economic Environment

The Economic environment additionally influences the patterns of tourist flow – especially from the tourist creating districts. Additionally, the conversion standard between two monetary forms can modify the expense of occasions accordingly an occasion bundle may become less expensive or more costly which likewise influences the tourist traffic.

2.7.3. Availability and Accommodation

Availability and Accommodation are the two fundamentals of tourism. Any adjustment of these influences the tourist traffic enormously. The adjustment of the recurrence of transport administrations or the start of any new administrations will undoubtedly influence tourist traffic alongside the distance and season of movement. Moreover, long stretch objections deal with an issue in such a manner. Truth be told a significant disadvantage for India in the American and European business sectors is the long stretch of travel to India from these tourist-creating areas. Essentially accessibility and paces of convenience will likewise influence the tourist stream.

2.7.4. New Advancements in the Business

New advancements in the business, for example, setting up of another fascination say an event congregation or some other amusement park will acquire more sightseers. Opening of new objections like "Sun City" in South Africa has attracted part of traffic from India. It is significant that the business players watch such changing patterns and rivalry to do determining.

2.7.4.1. Benefits of Interest Conjecture for Tourism

Request forecast in tourism is of extraordinary monetary worth both for the general population and private area. Any data concerning the future advancement of tourism streams is vital to hoteliers, visit administrators and different enterprises worried about tourism. In the most recent couple of many years, various analysts have considered worldwide tourism interest and a wide scope of the accessible interest estimating methods have been tried. Moreover, the significant center has been given to econometric investigations that include the utilization of least squares relapse. It is to assess the quantitative connection between tourism interest and its determinants. Be that as it may, econometric models generally neglect to beat straightforward time arrangement extrapolative models.

This article acquaints another methodology with tourism request determining by consolidating specialized examination procedures. The proposed model is assessed versus a scope of exemplary univariate time arrangement techniques as far as estimating and directional exactness.

2.7.5. Hazard Appraisal in Tourism

Despite endeavors embraced since the mid-twentieth century, the danger evaluation seen from the viewpoint of catastrophe has just been dealt with decently as of late. Its orderly origination and investigation were basically accepted by specialists in the normal sciences with considers in regard to seismic tremors, volcanic ejections, landslides, flooding and mechanical mishaps.

At the end of the day, accentuation was fixated on the information on risks because of the current insightful and scholastic inclinations and the endeavors of the individuals who previously considered these issues. It is therefore critical to call attention to here that the accentuation remains, especially in the profoundly evolved nations, where because of their mechanical improvement individuals attempt to discover more prominent detail the creating wonders of the dangers. Moreover, this was an apparent pattern during the main long stretches of the 'Global Decade of Natural Disaster Reduction' pronounced by the United Nations (UN) General Assembly.

Tourism ventures and those intrigued by their accomplishment in adding to the social and financial government assistance of a populace, need to diminish the odds that a choice will neglect to accomplish wanted destinations. One significant approach to lessen this danger is by knowing certain future occasions or conditions more unmistakably. Perhaps the main occasion is the interest figure for a travel industry item, be it acceptable, a help or a heap of administrations, for example, excursion or what an objective offers.

2.8. NEED FOR A DANGER EVALUATION DEVICE

All ventures are keen on such danger decrease. In any case, this need might be more intense in tourism ventures than for different businesses with different items, for the accompanying reasons: Tourism item is transitory. When an aircraft has taken off, or an amusement park has shut for the afternoon or morning first lights over a lodging, unsold seats, confirmations or dozing rooms evaporate, alongside the income opportunity related with them. This places a premium on forming request in the short run and expecting it over the long haul, to keep away from both unsold 'stock' from one perspective and unfulfilled interest on the other.

Individuals are indivisible from the creation utilization measure. Generally, the creation of tourism items happens simultaneously as its utilization. Furthermore, a lot of this creation utilization measure includes individuals interfacing as providers and customers, for example, inn staff, servers and servers, airline stewards and performers. Therefore, this puts a premium on having enough of the correct inventory faculty accessible when and where guests need them. Hence, consumer loyalty relies upon correlative administrations: The guest's experience relies upon fulfillment with a large group of labor and products offered by a lodging. Request figure of a lodging relies upon:

- the volume of aircraft flights and other vehicle admittance to its space,
- the nature of air terminal administrations,
- the neighborliness of cab drivers, and

- the quality and cost of diversion and the accessibility of sporting freedoms.

Nevertheless, request figure can help guarantee these correlative administrations are accessible when and where future guests need them, which will bounce back to the advantage of the lodging or other individual tourism office. Moreover, recreation tourism request is incredibly touchy to regular and human-made calamities. Many occasions and get-away travel are animated by the longing to look for shelter from the pressure of the regular climate. In addition, today there are incalculable choices for investing relaxation energy agreeably for inhabitants of most created countries. Therefore, emergencies like conflict, psychological oppressor assaults, illness episodes, wrongdoing and outrageous climate conditions can without much of a stretch prevent recreation explorers from visiting an objective experiencing one of these, or from going by any means.

On the other hand, the capacity to figure such occasions and their extended effect in tourism requests can help limit the antagonistic impacts of calamities on tourism related deals, pay, work and assessment income of a spot. Tourism supply requires enormous, long lead-time interests in plant, gear and foundation. Another inn may take three to a long time from idea to opening. Another air terminal or ski resort may require 10 years or so for all arranging, endorsements and development. Another plane may require five years to create from an aircraft's underlying request to conclusive conveyance. Future interest should be expected effectively if providers are to keep away from the monetary expenses of overabundance limit or the chance expenses of unfilled interest.

2.9. CONCLUSION

In the conclusion, this chapter discussed about the basic significance of tourism necessitates the forecasting. This chapter also discussed about the various kinds of forecasting techniques that have been used in tourism industry. This chapter also discussed about the several types of demand forecasting such as passive demand forecasting, active demand forecasting, long haul projections, external macro forecasting, and internal business forecasting.

In this chapter, different kinds of various strategies have been discussed that have been used in demand forecasting which includes trend projection, market research, salesforce composite, delphi technique, and econometric. Towards the end of the chapter, various types of factors that influences the tourism industry have been discussed as well. This chapter also discussed about the several kinds of social as well as safety factors such as geographical and climatic conditions, the economic environment, availability and accommodation, new advancements in the tourism industry, and hazard appraisal in tourism.

REFERENCES

1. Chetty, P., (2012). *Advantages of Demand Forecast for the Tourism Industry*. [online] Project Guru. Available at: https://www.projectguru.in/demand-forecast-tourism/ (accessed 8 August 2021).
2. Jake Rheude, (2020). Demand Forecasting: Types, Methods, and Examples | *Red Stag Fulfillment*. [online]. Available at: https://redstagfulfillment.com/what-is-demand-forecasting/ (accessed 8 August 2021).
3. Journals.elsevier.com. (n.d.). *Annals of Tourism Research*. [online] Available at: https://www.journals.elsevier.com/annals-of-tourism-research/curated-collections/curated-collection-tourism-demand-forecasting (accessed 8 August 2021).
4. Owlgen, (2020). What is Demand Forecasting? *Why Forecasting important in Tourism?* – Owlgen. [online] Owlgen. Available at: https://www.owlgen.in/what-is-demand-forecasting-why-forecasting-important-in-tourism/ (accessed 8 August 2021).
5. Song, H., Qiu, R., & Park, J., (2019). A review of research on tourism demand forecasting: Launching the Annals of Tourism Research Curated Collection on tourism demand forecasting. *Annals of Tourism Research,* [online] 75, pp. 338–362. Available at: https://www.sciencedirect.com/science/article/pii/S0160738318301312 (accessed 8 August 2021).
6. Study Moose. (2017). *Forecasting techniques in tourism demand Free Essay Example*. [online] Available at: https://studymoose.com/forecasting-techniques-in-tourism-demand-essay (accessed 8 August 2021).
7. Unit 6. *Forecasting for Tourism and Its Products*. (.n.d.). [ebook] Available at: https://hmhub.me/wp-content/uploads/2017/11/Unit-6-Forecasting-for-Tourism-and-its-Products.pdf (accessed 8 August 2021).
8. Viorica Ani Merlă and Sorin ChiriŃă, n.d. *Forecasting In Tourism – Important Component of the Planning Process*. [ebook] Available at: https://citeseerx.ist.psu.edu/viewdoc/download?doi=10.2.2.942.897&rep=rep1&type=pdf (accessed 8 August 2021).

9. Zhang, Y., Li, G., Muskat, B., & Law, R., (2020). Tourism Demand Forecasting: A Decomposed Deep Learning Approach. *Journal of Travel Research*, [online] p.004728752091952. Available at: https://journals.sagepub.com/doi/abs/10.1177/0047287520919522?journalCode=jtrb& (accessed 8 August 2021).

Chapter 3

Creativity and Innovation in Tourism Business

CONTENTS

In the chapter, creativity and innovation in tourism business, several innovations that have been undertaken by tourism companies are discussed. The chapter also highlighted current status of innovation in tourism companies. Some of the key technology trends emerging in the travel industry in 2021 are also demonstrated. It therefore explains the concept of creativity and its role and importance in tourism. It also shed some light on the innovation in tourism sector and smart tourism. At the end, some of the smart examples in tourism is explained.

3.1. INTRODUCTION

In year 2003, Porter and Ketels stated that innovation is considered as a major source of economic growth as well as competitive advantage. Innovation is a set of technological, scientific, organizational, commercial and financial activities aimed at commercialization of existing technologies, knowledge as well as equipment.

The outcome of innovation is additional or new service or product with newly developed features as well as qualities. There are a growing number of applications and research studied the phenomena of innovation where focus of business is mainly on the process of manufacturing sector. he researchers studied service Moreover, the sector related to manufacturing and some of them have considered the difficulties in applying theories of product development to services (Gallouj, 202 and Prajogo, 2006).

The innovation process is a process of development, creation, distribution as well as usage of innovation. In addition, the process of innovation in respect to the product is referred as a process of successive transformation of ideas into goods through the stages of applied and fundamental research, marketing, production, and engineering development.

Tourism is an important source of income and one of the major international activities for many countries. Therefore, the GDP of several countries is entirely generated by tourist traffic. The following feature drives the importance of tourism for the world economy. Tourism is a complex socio-economic phenomenon that is an economic activity. It is also a mean of inter-regional as well as international cultural exchange, a way to overcome several types of crises. It is therefore obvious that through this fast-growing industry there is an opportunity to understand innovation and define the different categories of innovation. On the contrary, Creative tourism is one of the forms of cultural tourism (Ohridska and Olson, 2010). More specifically, it is "travel directed toward an authentic and engaged experience, with participative learning in the heritage, arts or special character of a place, and it provides a connection with those who create this living culture and reside in this place" (UNESCO, 2006).

In year 2000, Richards and Raymond stated that it "offers visitors the chance to enhance their creative potential by active participation in courses as well as learning experiences which are characteristic of the holiday destination where they are undertaken."

According to Urry (1990), creative tourism goes beyond the general tourist gaze. It also involves deeply the tourists in the culture's cape of the destination where they take part in a number of activities such as crafts, arts, culinary and other creative activities.

Therefore, it creates a close link between the local population, the tourists, and its cultural heritage (Wilson and Richards, 2007). Now-a-days creative tourists are not satisfied with pure observation of cultural spectacles, but they also look for active participation. For instance, they travel to France to cook traditional Provencal dishes, to Argentina to learn to dance tango, to Russia to paint icons. Generally, creative tourism is perceived as a form of cultural tourism, but it is different from the mainstream cultural tourism and it will be further elaborated in the paper.

Traditional cultural tourism is based on "seeing," "viewing," and "contemplating" (e.g., visiting museums, concerts, art galleries, ballet performances and the like) whereas creative tourism is based on "experiencing" (Gilmore and Pine, 1999), "learning" as well as "participating" (e.g., not only observing icons or icon painting but taking courses in icon painting in the destination). With a primary focus of active skill development, it puts creative tourism as the next generation of cultural tourism that satisfies the higher-level need of self-actualization. Moreover, creative tourism is not so place bound as compared to cultural tourism due to the reason that creative tourism utilizes tourist resources that are processes in essence such as crafts, dances, painting, singing, festivals (Andersen and Prentice, 2003). Therefore, it is more sustainable in nature as compared to traditional cultural tourism that is based on the consumption of built environments (Wilson as well as Richards, 2006) and contributes to the development of the destination (Ritalahti, Lindroth, and Soisalon-Soininen, 2007).

3.2. INNOVATION IN TOURISM COMPANIES: WHERE ARE THEY?

Although innovation as a business issue caused by the interest of all types of businesses to make a competitive advantage dates back to 1934 with the proposals of economist Joseph Schumpeter, however it have re-emerged studies with greater momentum (Hartmann & Calderón, 2010). Moreover, tourism companies face an increasingly competitive as well as complex environment that forces them to change according to the environment and to use innovation as critical factor to the success and survival (Gálvez, 2011; Pérez & Gálvez, 2012).

Since the context of the tourism sector is marked by some peculiarities that prove to be more experienced, including the constant as well as sudden change of profile in the tourist, thus it becomes difficult to provide unique experiences with some degree of fascination and impact. Besides this, the tourist has increased as well as immediate access to information. Hence, tourist can organize his holiday independently while demanding tourist services sustainably managed in accordance with the social, environmental and economic dimensions, expressing his duty towards the environment (Gomezelj & Čivre, 2015).

Different angles to be applied are marked through the evolution of the concept of innovation, and at this, service companies focus on opportunities for adoption. We need to treat service companies differently since its characteristics operational, organizational and markets are different, despite recent attempts to make innovation. Thus, we can say that innovation in the services sector must be thought more widely as the complexity to provide a service be determined by ephemeral, intangible, extremely difficult to retain as well as changing attributes. In year 2002, Hjalager stated that given this, the term "innovation" has been widely used to explain the ways companies behave, destinations as well as tourism sector itself.

On the other hand, according to the viewpoint of Zbigniew (2015), innovation regarding tourism companies involve redirecting the traditional forms of management to those oriented towards innovation, i.e., in its forms of operation and organization seeking new ways to create new products, efficiency of the tangible and intangible resources and implementing powerful strategies of the company. Therefore, the concept of "tourism innovation" came into existence with the practice of innovation in the management of tourism enterprises, rooted conception that relates to the pragmatic side and the ontology of the attitudes of tourism and utility tourism production and consumption of innovative products (adapted or new). Nevertheless, Moufakkir and Dornan (2015) observed that innovation is a vast concept and includes simple adoption of goods and services for tourists as well as inventions. Hence, we can say that innovation is a strategic way of thinking oriented competitiveness together with value creation. Therefore, innovation in tourism includes walking towards a new management corresponding to increased competitiveness, value creation and development of unique collaborators capabilities, efficiency and resource efficacy, and chairing of leadership and innovation culture, not only at organizational level but also at market level.

In the same way, it is necessary to recognize the project as a mechanism that makes it possible to capture ideas as well as take benefits of them at the time they turn into innovations (García, Sánchez, & Mendoza, 2015). The company should focus on learning processes together with incremental improvements that lead to changes in organizational culture as well as participation in all areas of the organization in order to emancipate the development of innovation and lead to development in growth, productivity as well as competitiveness (Cruz and Puente, 2012; Sánchez et al., 2015). Therefore, the question to be resolved is to indicate where the companies in terms of innovation are. In simple words, discover as well as analyze what has been done in research on innovation in this sector. Hence, for countries in terms of competitiveness, thc tourism represents the entry of economic stability together with income and on the contrary knows where the tourism companies are going will help to get aware about future challenges for developing innovation in tourism.

3.3. INNOVATION IN TOURISM COMPANIES: CURRENT STATUS

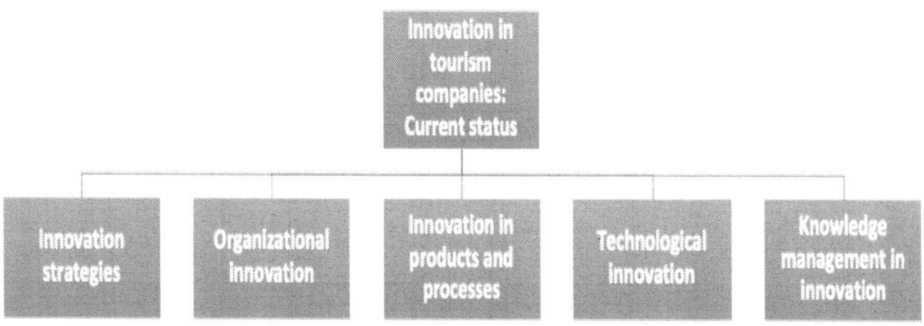

Figure 3.1. Innovation in tourism companies: Current status.

3.3.1. Organizational Innovation

Organizational innovation determines how a brilliant idea is put into practice in order to gain an advantage in the organizational structure i.e., new ideas align with the rethinking of organizational activities, so that the business improve their competitive level and boost its market value. It has been characterized by the application of knowledge to meet the needs of the companies (Souto, 2015).

In general, research on organizational innovation is related to aspects such as organizational culture, innovation capabilities, leadership, personality, enterpreneurship and creativity; as well as organizational barriers that are considered as limitations of the process of innovation.

Organizational aspects also are concerned with the membership as well as size of a group. Therefore, integrated and larger companies are considered those that come with a higher degree of formalization in its organizational routines and procedures, something that can be positive for innovation.

This entails harmonizing a number of factors providing members of the organization the opportunity to learn, involved with the internal operation, improve as well as contribute to the process of innovation.

In addition, the membership and size of a chain or corporate business enables companies to have better access to resources as well as capabilities, such as financial resources, qualified personnel, and access to information (Serrano, López, and Gómez, 2009; Vargas & Martínez, 2015).

In year 2011, Walsh, Lynch & Harrington observed that another approach to organizational innovation states that technology alone cannot provide benefit, thus needs to be managed to lead innovation capacities, these are strategically developed by dependencies organizational path that drives as well as stimulates the behavior in response to changes in context.

Among all these capabilities, absorption is identified and involves some level of transformation, assimilation as well as exploitation of resources; although its design must be re-evaluated according to the nature of the services in order to be implemented properly in the tourism sector (Thomas & Wood, 2014).

3.3.2. Innovation Strategies

The studies identified on strategies of innovation are classified into four sub-themes and that are market strategies, innovation policies, strategic alliances as well as pro-environmental strategies. Regarding to those who address market strategies, they have been able to visualize innovation as means to create value and competitive advantages in their products (goods or services), improve their positioning, attract new market niches, achieving internationalization, as well as avoid strategic drift.

Strategies of innovation play a major role in retaining customers and creating value in products as well as driving economic performance of the company. Therefore, it is desirable for every company to be engaged in strategic planning and development of innovations that are geared to fulfil the demands and needs of the market (Plank, Grissemann & Brunner, 2013; Santa & Nicolau, 2013).

Given this, González (2011) mentions that decision-making in tourism companies as well as open innovation processes are used as a facilitating tool in the entry and positioning in the tourism market; suggesting a system based on external agents and the management of organizational capital (structural, relational and human).

This type of system can promote the internationalization of a company, as strategy is to take action on the present situation, in simple words, be aware about the changes made by competitors and seek sources of knowledge to assist the reconfiguration of the innovation system in order to encourage participation in tourism markets globally (Shaw & Williams, 2011).

Over and above, innovation as market strategy is important to avoid strategic drift of a company that is caused when the company is unable to cope with the contingency of the environment, loses its own control and unable to continue its operations. Among all the aspects that counteract as well as prevent its effect are mentioned below:

- the proactive adoption of new technologies for new customer needs, creating products and expectations; and
- the culture of innovation is the basis for the generation of successful and creative solutions by staff to improve the quality of service and achieve cost reduction; thus, ensuring share of greater market (Edwards & Dwyer, 2009).

3.3.3. Innovation in Products and Processes

In year 2015, Albu stated that motive of innovation of products is to design and market launch of a new improved service or product to attract attention of consumer and whose novelty is apparent to the supplier or manufacturer as well as consumers.

On the contrary, process innovation is the production of particular changes in the internal orientation of the company, issues relating to the use of existing new and improved technologies, replaces an old production line with a new one and aspects that contribute to increase the performance of operations (Albu, 2015).

Despite the fact that the diffusion of innovation among small tourism companies has a low propensity for development of new improved products as well as processes (Moreno & Tejada, 2013), the research conducted in this line converge on innovation in products and processes is more frequent in restaurant companies as compared to other companies.

This is due to its hybrid nature of operation, i.e., is comprises of two parts, one operative similar to industrial companies, and other, of service where the customer is involved in the process (Ottenbacher & Harrington, 2009; Albors et al., 2013; Pellegrin & Leszczyńska, 2013; Tüzünkan & Albayrak, 2015).

An important part of the tourism system is the restaurants. It attracts customers and boosts the competitiveness of the sector. Therefore, it is necessary to direct product innovation towards customer satisfaction by the developments in food services and application of cooking techniques.

For instance, useful applications of molecular cuisine are to innovate dishes, impacting positively on the pleasure of customers and perception of quality in restaurants (Tüzünkan & Albayrak, 2015), as well as creating new dishes based on contemporary or traditional cuisine, special minimalist designs as well as alternative ingredients like vegetable-based gels, sugar substitutes, etc. (Albors et al., 2013).

Demonstrating that innovation in culinary products increases profitability at the same time generating an attraction (Albayrak & Tüzünkan, 2015); also, the culinary product cannot be separated from the process of innovation, as it is the outcome of this and the complex assembly of the supply chain (Leszczyńska & Pellegrin, 2013).

For the tourism sector, the rise in the culinary supply improves food quality, the efficiency of operations (kitchen), flexibility of food preparation, compliance with regulations and standards food service, service quality dining, reduces losses and increases customer (Albors et al., 2013). Hence, we can say that it is not surprising to talk of culinary innovation, a concept that has been developed by innovation leaders as Ferran Adrià or Michelin star chefs.

This innovation has been defined as an experimental process that involves giving an idea that initially probably is not well structured, but with the passage of time, learning, teamwork as well as dissemination is strengthened.

The objective of this new idea is to solve the problems. In year 2013, Albors studied that innovation drives produced from a kitchen transforms the tacit knowledge into unique competitive advantages for chefs.

3.3.4. Technological Innovation

The technological innovation where the Information and Communication Technology (ICT) is aimed to the promotion, distribution as well as marketing of tourism products is another topic addressed by technological innovation researches.

According to Puccio & Grana (2008), such studies are future tools for competitiveness in a tourism market that will require the intensive use of technologies. Generally, the tourism sector has been adopting gradually technology to enhance the interactivity of companies with their stakeholders and customers.

This is because more and more people are using the ICT (including Global Distribution Systems [GDS], Computerized Booking Systems [CBS] and Internet) to acquire different tourism products (Pansiri & Courvisanos, 2010). ICT offers the following advantages:

- possibility of search and add products to the basket (dining booking online, accommodation payments, among others;
- capacity to set a fixed price for the package as per the discount regulations, and
- wide access to suppliers and products;
- security while paying electronically for a package tour (Găzdac, 2009).

Over and above, it has been shown that the use of different ICT has a few effects in the increase of productivity as well as on the level of competition. In the meantime, they have favorable effects on increasing the market share of the companies because it promotes the releasing of new products and in the improvement or introduction of new processes.

Therefore, its ability to lead local businesses to global markets is potential, (Ollo & Aramendia, 2013). ICT have a wider use in big companies and because of this, the innovation structures are far-reaching and more formal, also they stimulate human resource management and the work practices. Moreover, through the use of the Internet, the product offering is radically innovative, for operating activities and marketing that benefit the increasing of earnings and the resources optimization, especially the human, that encourage the collaboration to innovate (Ficapal, Torrent, Vigil & Boada, 2015; Neykova & Miltchev, 2015).

3.3.5. Knowledge Management in Innovation

In companies, knowledge as well as its management is perceived as a strategic asset that offers a number of competitive advantages. In year 2014, Yiu & Law argued that tourism is a knowledge based because of the latest development of information, production, processing, exchange as well as transfer of knowledge that have brought effects in their processes of organization, operation together with interaction.

Nevertheless, innovation does not rely on the amount of knowledge that go around, but the use made of it so that it can allow high levels of competitiveness (Alabart and Ponce, 2015). Various other researches take elements of knowledge and information that come from technology, customers, and personnel together with the environment.

Thus, we can say that knowledge management refers to the use of knowledge as well as information to become capacities for competitiveness and innovation (Shaw & Williams, 2009; Tonelli, Simaro & Carús, 2012; Nieves, Quintana & Osorio, 2014). The different types of innovation (process, product and marketing) are influenced by the level of knowledge managed inside the organization, external social relations as well as efforts to collaborate with people. In simple words, companies focus their efforts in favor of assimilation and sheltered knowledge so that they can turn it into capital (Nieves et al., 2014). It has been recognized that brainpower is one of the elements involved in innovation and that can be obtained by the proper management of knowledge. Its most important function is to create value for the companies related to tourism (Monagas, 2012).

The human that is intangible, hidden, valuable and difficult to imitate is one of these capitals. Moreover, it makes a significant contribution in achieving innovation in non-intensive service companies in hard technologies (Rivero & Rastrollo, 2013; Nieves et al., 2014). By investing in human capital and establishing systems that enable the integration of knowledge of the different members of the organization, companies can be more successful in developing innovations.

3.4. KEY TECHNOLOGY TRENDS EMERGING IN THE TRAVEL INDUSTRY IN 2021

A crucial part of the tourism and travel industry is technology. It helps businesses with day-to-day operations and at the same time improves the experiences of customers. For this reason, it is important that airlines, restaurants, hotels as well as other companies keep themselves up to date with the latest tech trends within the travel industry. This is especially necessary in the scenario of COVID where expectation of customers keep shifting. You can find out more about some of these new technology trends for 2021 in this article. We piled up the newest as well as the latest technology trends in travel and tourism industry in addition to tech trends that serve as an answer to bring the change in consumer behavior because of the COVID virus.

3.4.1. Voice Search and Voice Control

As far as technology trends in travel industry are concerned, smart speakers, smartphones and AI assistants have all helped to increase the relevance of voice search. In general, many travel customers use voice search for several purposes such as find as well as book hotel rooms, airline tickets, travel experiences, etc. Therefore, we can conclude that it is very important to keep voice search in mind while designing your website so that you can capitalize on this trend. Over and above that, voice control is playing a major role in actual travel experiences too. Voice-controlled devices can be used inside the hotel rooms to obtain tourist information, or to control heating and lightning within rooms without needing to speak to a member of staff. It is expected that this will continue to expand into other areas of the industry too.

3.4.3. Robots

One of the most exciting forms of travel technology is robotics technology and it is constantly improving. This technology is used within hotels to serve various purposes, for example, robots have been used in concierge-like roles, helping to greet guests when they arrive and provide information. A few hotels have expanded their use and involving them in luggage handling and cleaning. Moreover, robots can potentially have a role in food preparation as well as food service in restaurants. They can also be used in airports to detect concealed weapons. Moreover, some manufacturers also use robotics in order to create luggage cases that intelligently follow you and travel agents are using the same for pre-screening, so that they can make waiting times more productive for customers. Moreover, along with various other technology trends within the travel industry, the uses for robots have increased in response to COVID due to the reason that they have the potential to reduce human-to-human contact.

3.4.3. Contactless Payments

The ability to accept contactless payments is another important form of travel tech. This will help travel companies to process payments quickly, including situations where customers do not have access to their credit or debit card and cash. As it is a time saver, therefore, customer experience can also be enhanced. Moreover, the system of contactless payments was introduced for making the payments convenient. However, in the wake of the coronavirus pandemic, it has emerged as a key part of travel marketing. Travelers feel reluctant to handle cash, as they are nervous about the spread of the virus. You can protect your own staff by using contactless payments as much as possible.

3.4.4. Virtual Reality (VR)

Figure 3.2. VR in tourism.

Source: Image by Business World IT.

The role of virtual reality in tourism industry is especially significant and it has been an emerging technology in several different sectors. It allows travelers to experience locations at distance from the comfort of their own home and it can therefore be the difference in whether they eventually complete a booking. With the usage of VR tours, customers can experience everything from virtual hotel tours as well as restaurants, to national parks, landmarks or even particular activities. The immersion and interactivity can also help to provide a competitive advantage over rivals who are unable to make the most of these types of tech trends. Moreover, by now, the most 360 tours or VR tours are compatible with mainstream web browsers. it can also assist those who feel reluctant to travel because of COVID to look at the benefits more clearly.

3.4.5. AI Chatbots

AI-powered chatbots are most sensible travel technology investments that one can make. This is due to the reason that these chatbots can provide customers with swift answers to their questions on a 24/7 basis, regardless availability of staff. This can be necessary for fulfilling current expectations related to customer service. In most cases, these chatbots are considered the best at the time of answering common questions that have standard answers. Moreover, significant progress is made by conversational AI and its adjacent technologies to finally understand the mysteries of human language, speech, communication, emotions and real intent.

This growth in conversational AI is just the beginning, and the future is going to see the deployment of conversational AI in home automation, automobiles, marketing, advertising, e learning as economies globally undergo digital transformation. They can also be especially useful while responding to queries related to COVID policies and hygiene or safety measures.

3.4.6. Cyber Security Measures

In recent times, tourism and travel companies have increasingly come on the radar of hackers and cyber criminals who are out to exploit vulnerabilities associated with data security data breaches have become a common thing nowadays in the tourism and travel industry. Moreover, travel companies are a major target due to the reason that they employ many people and have access to an abundance of data related to customers. Moreover, some of the biggest threats in this area involve ransom ware attacks and phishing attacks; however, the modern reliance on data also leaves companies at risk from human error due to their own employees too. This requires investment in cyber security training as well as various software and hardware solutions in order to keep your business safe. Finally, is therefore important to comply with data protection laws and to keep up with the latest regulations.

3.4.7. Internet of Things (IoT)

Figure 3.3. IoT in tourism.

Source: Image by Tommorow.city.

The Internet of Things (IoT) is one of the most exciting emerging technology trends. The IoT has the potential to have huge impact businesses, automating processes without any human-to-computer or human-to-human interaction. Based on the sheer volume of adoption and the ability to offer advanced interconnection and communication among devices, systems and services, the Internet of Things is expected to bring disruption in the travel industry. IoT is streamlining the end operations of the hotels, airlines and other travel companies by connecting smart devices, systems and processes. Hence, by taking advantage of the IoT technology, travel industry can realize increased operational efficiency and more personalized guest experience. For example, in hotel rooms, IoT technology can be used to provide customers with a device that connects to everything from the heaters to the lights as well as air conditioning, allowing all to be controlled from one place. Meanwhile, in airports, luggage cases are installed with sensors that alert passengers when they pass by.

3.4.8. Recognition Technology

Recognition technology in the tourism industry consists of outstanding benefits to the user, including enhance more security, service nontransferability, and embrace a high level of efficiency. The technology itself includes retina scanning, fingerprint recognition, facial recognition as well as various other biometric identifiers. Moreover, in some hotels, such technology is already being used to allow access to rooms via fingerprints. Hence, customer will be able to pay for meals in the hotel restaurants simply by walking through the exit.

3.4.9. Augmented Reality (AR)

Augmented reality is just like virtual reality. However, it involves augmenting real surroundings of a person instead of replacing them. It is cheaper than VR and this is one of the major plus points of this technological trend. Users are required only a tablet or smartphone device along with access to the internet. Moreover, AR based travel apps enhance the customer experience; provide customers with valuable information or even pure entertainment. For example, apps can allow photographs to be augmented by using effects and filters. It makes possible to display the details about local destinations as a customer points their smartphone at them, providing information at the exact time that it is most relevant.

3.4.10. Artificial Intelligence (AI)

Artificial intelligence is improving all the time, making it more reliable and more attractive as a business solution. Therefore, businesses in the travel industry are taking advantage of AI in order to perform a variety of administrative and customer service tasks.

Figure 3.4. Application of artificial intelligence.

Source: Image by Wikimedia commons.

The concept is strongly linked to ideas of automation, where processes are carried out with little or no human intervention. AI can save businesses time and money, while potentially eliminating human error and allowing tasks to be performed quickly, at any time of the day. Therefore, most hotels and resorts rely heavily on delivering excellent customer service to build their reputation and AI technology can assist with this in a wide variety of different ways. For example, artificial intelligence can be used to improve personalization, tailor recommendations and guarantee fast responses times, even in absence of staff.

3.4.11. Big Data

Big data plays a very crucial role in modern tourism management and almost all successful companies employ their own techniques of data collection. The biggest use for this data is to improve personalization with travel companies with the usage of information that they gather in order to make adjustments to their offerings. In order to analyze current business performance is another valuable use of data. Owners of hotels can use big data for purpose of revenue management with usage of historic occupancy rates as well as other past trends to anticipate levels of demand in a better way. Pricing and promotional strategies can also be optimized when demand is predictable.

3.4.12. 5G

Travel technology has become more powerful with help of 5G networks. They promise wider coverage, more stable connections as well as much faster loading and downloading speeds. 5G allows us to develop and deploy technology that 4G limited us. It also allows downloading the content 20 times faster than before. In other words, the connection between smart devices will be more efficient. In addition, we will able to start truly enjoying the Internet of Things (IoT).

Experiential travel also known as immersive travel, where technology turns travelers into the experience's protagonist will be able to become a reality. Moreover, 360° or augmented reality (AR) video will be accessible as well as ubiquitous. Moreover, at the Roman Baths in Bath, England, the BBC ran a test project that used 5G and an AR application where users could go back in time to reconstructions of the site in key moments throughout history. One of the videos shows the pilot testing, which observed that over

80% of participants reporting would be more willing to visit a museum in case it had experiences related to that caliber.

3.5. WHAT IS CREATIVITY?

Figure 3.5. Creativity in tourism business.

Source: Image by Pngtree.

Creativity is used in the different dimensions and it has a multidimensional approach. It can range from technology-driven creativity (creation of online models of tourism), economic creativity (enterpreneurship) or last but not the least cultural creativity (creativity in showing culture, heritage of the place).

Creativity in tourism can lead to creation of creative economies using a tourism model, creative policy, creative environment, creative tourism academics and creative villages and last but not the least, creative tourism experience. In tourism, understanding the concept of creativity is the analytical review of value chain analysis creating creative nations for tourism.

3.5.1. Channeling Creativity in Tourist Experience

In tourism, experience of products can be seen in different ways because of features of its products as well as the reasons underlying behind the consumption of these products. In year 1987, IsoAhola and Mannel in his work clearly identified the main indictors for the consumption of tourism products as well as push and pull factors for tourists.

The work focused on dimensions of consumption of tourism products, leaving behind of personal as well as interpersonal problems of their environment. Providers of tourism products are trying to become more creative as well as tourists themselves are becoming more creative by demanding creative products which in turn helps in providing one of the most unforgettable experiences from the trip.

Here, the word trip has been used in the context of experience that is gained from the source to the destination and the other way around. One recommends the destination for future when tourists get a better tourism experience.

It can be achieved by removing the gaps in the tourism industry. The main ones in the service industry with intangible nature of tourism products such as the service design gap, the communication gap, the listening gap and the service delivery gap. In overall tourist's experiences, creativity at any one-sub component of delivering a creative tourist product can be a great achiever in channeling creativity.

Today's modern tourists with better knowledge through involvement in consumption pattern and multi variety of sources have led to the creation of creative experiences economies of tourism. The tourist has become comparatively more conscious as well as systematic in the approach of consumption of tourism products of his choice.

Understanding the concept of experience economies in the service industry allows the service provider (Government bodies as well as private bodies) to be creative in delivering a different and better tourism product.

In order to provide an unforgettable experience to the tourist, it helps to develop methods of his or her expertise and channel its energy in the right direction.

Thus, there is a requirement to create a wonderful hardware, an effort that can be at both micro and macro level. Hardware refers to infrastructure that enables tourist to have an exclusive experience at the time of traveling to a destination and software that is developing of experiences at the destination. In the region, this will lead to creative experience economies.

A creative experience economy is to be created at both the rural areas as well as the urban areas of the country in order to make creative agglomeration of tourism infrastructure. Alfred Marshall, an economist of industry cluster, introduced the idea of creative infrastructure for tourism.

Industry clusters for tourism can be particularly identified tourism zones in rural and urban areas helping in improving competition and cooperation among economic units, specialized financial institutions, and specialized suppliers of tourism products together with other main bodies in the development of tourism in the region.

Channeling creativity in tourist experience also involves more spending of tourist at the destination with a variety of different tourism products that other suppliers of tourism products or destinations are not offering.

In year 2000, Gupta and Valij introduced a concept of creating a creative experienced tourist in the form of managed tourist. Creating a creative destination is a combination of challenges taken by a tourist in experiencing something unique and the challenges taken by several other bodies becoming creative host with creative accommodation and creative infrastructure at the destination.

It brings equilibrium in the supply as well as demand of products related to tourism. A good example in creative tourism is adventure tourism, as it has no limits to adventure as an activity and tourism as a concept. In channeling creativity in the adventure tourism domain is the foremost effort, enthusiasm, together with involvement by the individual in having out of box experience. In creating new form of creative tourism entities, concepts of prestige motivation can be used.

This concept can further develop creative suppliers of tourism products as well as creative potential tourists like some enthusiastic tourism lovers and suppliers are accepting the idea of space tourism. In the last, it can be concluded that creating a creative destination involves the challenges taken by the tourist and the challenges to which a destination comes up to attract a quality tourist to a destination, however, it asks for a balance between both the parties (Scitovsksy, 1976).

3.6. INNOVATION IN TOURISM SECTOR AND SMART TOURISM

In the tourism sector, different companies like travel agencies, hotels, car rental companies, restaurants and entertainment companies provide a variety of services. The tourism sector is a growing as well as demand-driven sector (OECD, 2006: 33). However, tourism is not just about a business that offers tourism products for tourists participating in tourism activities and demanding the services of these companies. Therefore, the socio-cultural structure of the local people and psychology as well as habits of the tourists affects the tourism experience because of the abstractness of tourism products. Moreover, in order to satisfy the tourists and ensure the continuity of the tourists, companies in the sector are in competition with each other. For this reason, there is requirement of innovation to gain a competitive advantage.

Another reason for innovation-oriented tourism policies is to encourage growth and increase productivity. This can only become possible when the customer is integrated as well as satisfied with it. Therefore, individuals participating in mass tourism provide their knowledge by sharing their experiences. Moreover, in tourism, innovations in information and communication technology enable enterprises to analyze this information from customers more efficiently and effectively (OECD, 2006: 58). This allows the creation of databases for innovations that are required to be done in the field of process, product, organization or market.

Nevertheless, the tourism industry is unable to consider independent and separate from the technology in this context. In tourism industry, technology enters via Global Distribution Systems (GDS), Computer Reservations Systems (CRS), Internet and Smart Technologies (connected through sensors and data clouds) (Koo et al.,2015:99). Moreover, the concept of Smart is used in economic, social as well as technological developments that are nourished through technological infrastructures. In these developments, the most important indication of addressing about the concept of smart is that open data, massive data, technological perceptions, new ways of connection as well as knowledge exchange are being used in the developments. Hence, it involves advantages in communication together with synchronization between institutions or individuals instead of individual benefits. This is because smart systems make the world a small village, provide data integration and share, accelerate communication and interpret as well as optimize complex analytical models (Gretzel, et al., 2015: 179). Thus, it becomes possible to make accurate operational decisions.

On the other hand, "Smart" technology, which we are constantly intertwined with applications like smart cards and smartphones in everyday life, is supported by several different physical infrastructures as well as confronts with examples such as intelligent factory, intelligent home etc. Moreover, the concept of "smart city" which is equipped with modern technologies as a pillar of innovation, fairly and effectively managed even within the tourism industry and without consuming resources in the concept of sustainability.

On the other hand, in order to increase resource efficiency as well as reduce consumption, smart cities use information and communication technologies at the highest level (Gretzel, et all., 2015:559). These innovations result in increasing the quality of life for the city dwellers that is the local people. (Lopez de Avila, 2015:24). Smart cities benefit from information and communication technologies (Hojer and Wangel, 2014: 7), which are digital in the dimensions of smart people, smart living, smart governance, smart mobility, smart environment as well as smart economy. Moreover, the concept of smart city also involves multiple industries, including tourism.

Smart tourism is a practical attempt and an important part of the smart city strategy (Guo, 2014:55). In fact, these dimensions can also be applied directly in tourism destinations. Therefore, it can be concluded that smart city concept gives origin to smart tourism concept. The UNWTO underlines the links between new tourism destinations and modern tourism development (UNWTO). It is hence very difficult to recognize modern tourism development out of the innovations of the modern world, where information and communication technologies are heavily used.

In this connection, Gretzel et al. (2015) emphasizes that both telecommunications as well as information in the tourism sector are mobile and modernized by integrating with social media adaptation, global distribution channels, web-based technologies and central reservation systems. New tourism destinations refer to a concept that has been created in the smart tourism of the smart city concept under the name of smart destination. What is meant by smart tourism? Although there is not a unique definition for smart tourism, in addition to tourism intelligence, intelligent tourism and e-tourism, the combination of tourism as well as ICT is becoming popular about the incorporation of mobile internet or social media with tourism (Chao, 2015).

Nowadays, smart tourism is a major contributor in the development of the tourism industry in the wake of the mobile technology is in the hands of potential tourists with opportunities to make online bookings, restaurants, maps or online bookings, to catch campaigns and to buy airplane tickets. Individuals can see, know as well as choose their tourism alternatives in the absence of losing money and time. As per the views of Hunter, et all., (2015:110), smart tourism makes information more reflective as well as more interactive of the interests of the user. It therefore makes new resources and capabilities available to the traveler as well as to destination-based stakeholders.

3.7. SMART EXAMPLES IN TOURISM

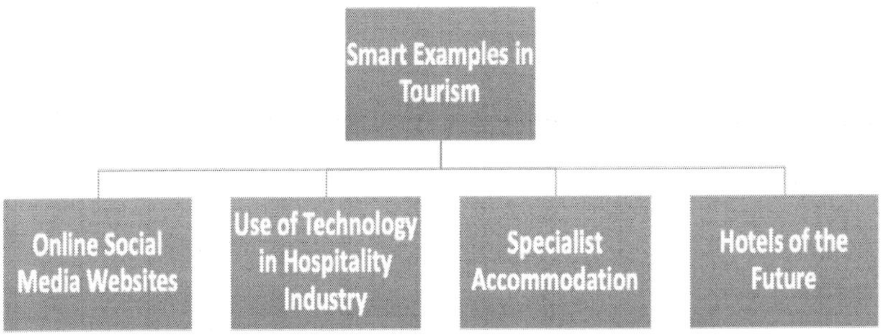

Figure 3.6. Smart examples in tourism.

Enterprises are continuously looking for ways by which they can upgrade the quality of their products and services that can lead to success. Tangible products involve goods such as curtains, hairdryers or alarm clocks in hotel rooms and services that are intangible in nature involve facilities such as the period of the cleaning of the room or the speed of internet in the hotel. Therefore, it is necessary to improve these products and services from time to time or replace with innovations within the scope of technological changes for the sake of satisfaction of the customers. In particular, the feedback from customers is considered as an important resource for these developments. Moreover, the other way should be keeping the eyes open for following new trends, challenges, learning new technology as well as products (Minett and O'Shannessy, 2008). In fact, enterprises want to neither remain behind the competition nor lose the customer. Nowadays, in the smart tourism gateways, new products and services are being developed in consideration of technological developments as well as innovations. A few of these are as follows.

3.7.1. Online Social Media Websites (Such As Facebook, Booking.com)

Intermediaries or travel agencies had chosen from a limited number of hotels selected from catalogues in the past whereas nowadays airplane, hotel and restaurant reservations can be made anywhere in the world with the usage of social media channels. Potential customers as well as customers can easily find places within the price ranges, they choose from the regions in the standard they want. In the tourism sector, increasing social media channels enable to make booking online as well as track positive or negative consumer comments with the implementation of Smart systems. Therefore, ideas are formed in minds of consumers even for small hotels that are not recognized and are not included in the famous chains and these hotels can be translated on travel. Moreover, these hotels can be recommended for traveling purposes (Seegers and Vermeulen, 2009:123–124). In the given case, we can conclude that the inclusion of online technologies has left behind the best understanding of traditional 5-star hotels.

3.7.2. Use of Technology in Hospitality Industry (Such As Tablets in Guest Rooms and Restaurants, Smartphone Keys)

Product innovation is one of the types of innovation as it is documented in the first part of the study. In these types of innovations, significant improvements are made to the existing products in order to provide improvements in terms of speed and efficiency. Methods used in process innovations that are other types of innovation are improving. Moreover, the objectives as well as starting points of such innovations in goods and services should be to increase satisfaction of customers. Hence, it is a fact that increasing customer satisfaction will bring loyalty, continuity and competitive advantage to enterprises. In this context, the technological innovations that can be used or are being used in hospitality enterprises are confronted as improvements in processes as well as products. For instance, these days people often travel with their smartphones, tablets or/and computers.

These people need internet service in their accommodation during their travels. Earlier many hotels were offering internet services at certain rates per hour but now they are starting to give this service free of cost in various parts of the hotel such as restaurants, lobby, guest room, fitness center, etc. with the awareness of their needs. Therefore, in some hotels, smartphones also now function as room keys. Hotels will install systems of smart room access at large scale that will enable guests to unlock their doors through simply swiping their phones across a keyless pad on the door. Moreover, Starwood who is owner of the Sheraton, Weston and "W" hotel chains has already installed 30,000 room locks across 150 hotels with this system. Moreover, Hilton will be implementing a similar system soon (Aurecon group).

Guests will not lose their keys and front office workers will not be involved in the key exchange process with the usage of this system. Another example of innovation in hotel rooms; the Aria Resort together with Casino hotel located in Las Vegas enables hotel guests to control the radios, air conditioner, lights, curtains, televisions in a room on a tablet. Moreover, the tablet can turn on and off the "Please do not disturb" light outside the room door and provides charging via USB ports.

3.7.3. Specialist Accommodation (e.g., Capsule Bed Hotels)

Today, one of the innovations in hospitality is accommodation in the capsule that was first developed in Japan in the 1970s (Kotaku) and then spread to the world. These hotels are also referred as Pol hotels. Capsule hotels are considered as one of the reasonably priced accommodation hotels. In these types of hotels, the guests stay in box-like compartments where they can only lie down and sleep. It is also possible to find these types of hotels in Amsterdam that have become a trend and was started in Japan and spread rapidly in other countries of the world. In Figure 3.6, Amsterdam City Hub offers wireless internet access in each area, compartments reasonably priced and a panel of light as well as a window for claustrophobics as compared to several hostels (Business insider).

3.7.4. Hotels of the Future (e.g., Space Hotels and Robotic Hotels)

These days there are some hotels where robots are used. For example, these robots pick up and place luggage at the Yotel hotel in New York, bring towels to guests in the pool at the Aloft Cupertino hotel in California and make check-in at the Weird Hotel in Japan. It is therefore anticipated that robots will clean the rooms and technological innovations will reduce the operating costs of the hotels in the future (Rentokil). Moreover, those who predict the possibility of space tourism in the future develop the concept of the spaceship. It is predicted that the space hotels that can be built in space would move at 27,500 km/h (17,000 mph) and circling the planet every 90 minutes.

Nevertheless, 16 sunsets as well as sunrises would be visible (BBC). Hotels in orbit will offer the services given like in other hotels, meals, bars, private rooms. Hence, they will also offer two unique experiences and that are stupendous views of the Earth as well as space and the endless entertainment of living in zero G involving further possibilities such as space walking (Space future).

3.8. CONCLUSION

At the end, it is concluded that creativity and innovation is much needed fuel that help in providing wings to tourism sector. This is one of the most important segments that helps in enticing customers and make them choose one destination over another. Therefore, it is often seen that countries that employs innovation and creativity into their tourism business grow significantly, resulting in increasing their revenue, foreign reserves, and employment generation. In addition, it also allows countries to improve the standard of living of their own country. For example, Dubai invested enormously in tourism business, resulting in surging their tourism base. It is continuous process that is ever lasting because customer needs and wants are continuously evolving and it is essential to cope with the same.

REFERENCES

1. Bárbara, N., (2021). *The Business of Creative Tourism and Creativity in the Tourism Business*. [online] Academia.edu. Available at: https://www.academia.edu/29983051/The_Business_of_Creative_Tourism_and_Creativity_in_the_Tourism_Business (accessed 8 August 2021).
2. Bascavusoglu-Moreau, E., Kopera, S., & Wszendybył-Skulska, E., (2013). The Role of Creativity in Development of Innovation in Tourism. *Journal of Entrepreneurship, Management and Innovation*, *9*(1), 5–15.
3. Bhatia, A., (2016). *Channeling Creativity into Competitiveness through Value Chain Analysis in Tourism*. [ebook] Amity Research Journal of Tourism, Aviation and Hospitality. Available at: https://amity.edu/arjtah/pdf/vol1–2/1.pdf (accessed 8 August 2021).
4. Delgado Cruz, A., Vargas Martínez, E., Montes Hincapié, J., & Rodríguez Torres, F., (2021). *Innovation in Tourism Companies, Where Are They and Where Are They Going? An Approach to the State of Knowledge*.
5. Hjalager, A., (n.d.). Strategic innovation in tourism business. *Strategic Management in Tourism*, pp. 102–117.
6. Insights. (2021). *Building the Future of Tourism Through Innovation – Insights*. [online] Available at: https://insights.ehotelier.com/insights/2015/04/20/building-the-future-of-tourism-through-innovation/ (accessed 8 August 2021).
7. Lim, W., (2016). Creativity and sustainability in hospitality and tourism. *Tourism Management Perspectives*, *18*, 161–167.
8. Malik, Y., (2021). *Importance of Innovation in the Travel Industry in the Covid-19 Era*. [online] YourStory.com. Available at: https://yourstory.com/2020/09/importance-innovation-travel-industry-post-covid (accessed 8 August 2021).
9. Moreau, E., Kopera, S., & Wszendybył-Skulska, E., (2021). *The Role of Creativity in Development of Innovation in Tourism*. [online] jemi.edu.pl. Available at: https://jemi.edu.pl/vol-9-issue-1–2013/the-role-of-creativity-in-development-of-innovation-in-tourismElif Bascavusoglu-Moreau (accessed 8 August 2021).
10. Perederenko, K., (2021). *Creativity In Tourism For Business Competitiveness*. [online] Academia.edu. Available at: https://www.

academia.edu/23505760/Creativity_in_tourism_for_business_competitiveness (accessed 8 August 2021).

11. Researchgate. (2021). *Creative Tourism Business Model and Its Application in Bulgaria*. [online] Available at: https://www.researchgate.net/publication/228257477_Creative_Tourism_Business_Model_and_its_Application_in_Bulgaria (accessed 8 August 2021).
12. Revfine.com. (2021). *11 Key Technology Trends Emerging in the Travel Industry in 2021.* [online] Available at: https://www.revfine.com/technology-trends-travel-industry/ (accessed 8 August 2021).
13. Richards, G., (2020). Designing creative places: The role of creative tourism. *Annals of Tourism Research*, 85, p.102923.
14. · Demi Roxane Horrocks, 2015. Creative Tourism: A Transformational Experience?. [ebook] Netherlands: Wageningen University. Available at: <https://edepot.wur.nl/354204> [Accessed 13 August 2021].
15. Teodorescu, N., Stăncioiu, A., Sidonia Răvar, A., & Botoş, A., (2021). *Creativity and Innovation – Sources of Competitive Advantage in the Value Chain of Tourism Enterprises*. [ebook] Available at: http://store.ectap.ro/articole/1055.pdf (accessed 8 August 2021).
16. TravelDailyNews International. (2021). *IFITTTalk on "New Technologies and Innovation in Tourism."* [online] Available at: https://www.traveldailynews.com/post/ifitttalk-on-new-technologies-and-innovation-in-tourism (accessed 8 August 2021).
17. Uhuru Network. (2021). *Tourism marketing: 11 Creative Ideas to Market Your Destination*. [online] Available at: https://uhurunetwork.com/tourism-marketing/ (accessed 8 August 2021).
18. Vidal, B., (2019). *The New Technology and Travel Revolution*. [online] Wearemarketing.com. Available at: https://www.wearemarketing.com/blog/tourism-and-technology-how-tech-is-revolutionizing-travel.html (accessed 8 August 2021).

Chapter 4

Strategic Management in Tourism

CONTENTS

This chapter explains the basic significance behind the ideology and concept of strategic management in tourism. This chapter also explains the various components of the strategic management process such as environmental scanning, strategy evaluation, and so on. This chapter also provide highlights on the several essential tasks of strategic management. It therefore addresses the several numbers of principles to successful and effective strategic management such as communication, leverage strength, etc. This chapter also sheds light on the strategic objectives and level of strategy. This chapter emphasizes the role of hospitality and tourism strategic planning in tourism sector, and with the importance of strategic management during a hospitality business.

4.1. INTRODUCTION

Promotion in the world of new and upcoming tourism has no denying that there are some unique tourism plans, especially around the advent and advancement of tourism development strategies. The implementation of tourism development strategies has been identified through the role of strategic management strategies in the tourism sector. Strategic management is an active process aiming to achieve long-term stability in the integrated planned tourism environment. Moreover, this management is a useful way to implement major development policies in the tourism sector set by the national economy which is influenced by tourism development. Hence, strategic management is covering essentially all the qualities required that promise quality and effectiveness in order to achieve the development goals set in the tourism industry.

Figure 4.1. Strategic management in tourism.

With intensifying competition and global homogeneity, the motivation and development of the tourism industry, the implementation of strategies to ensure efficient business operations (resource management, data management system, conservation and development of tourism, expansion and development of the tourism business), management of new modern tourism projects are all flourishing. Therefore, it is integral to strategize and manage in the absence of a successful labor sector that threatens the development of the tourism sector. Hence, strategic management is often taught as an important topic at some stage of a tourism degree program (usually during the last year).

Nevertheless, strategic management serves as a key function of the degree program. This topic is based on what students have learned in many years of research and applies this knowledge to the tourism environment at a practical level. However, many of the concepts associated with strategy and management strategies are trivial and often difficult for students as well as teachers to understand. It is therefore important to have access to resources that enable teachers and students to better understand the subject and create opportunities to apply the theory effectively. There are many modern resources, but these are related to the flourishing industry and it can be difficult to find a specific resource that focuses on the essence of tourism.

In fact, there are still many strategic planning manuals which focus on tourism. Therefore, this guide is very useful for those who teach management skills that apply to tourism students. This guide contains examples of generosity to help instructors teaching management strategies with hospitality. Moreover, the road guide is divided into several sections. Descriptive information about the toolkit is displayed first. This section is divided into major topics that may include semester levels in management strategy. It must therefore be noted that the framework and resources presented reflect this approach, while the facilitator uses the traditional framework in the analysis of management strategies. As the two main textbooks on management and tourism strategy, this guide uses Evans et al. (2003) and Tribe (1997) as specific applications of management theory in tourism. He led teachers into literature. In addition, a series of textbooks on strategic management strategy (lots of them!) are used to guide readers to chapters and case studies on related topics.

The guide also contains several related articles related to strategic management, as well as a tourism magazine, which usually contains essays and research papers on strategic issues. In addition, there are many online resources that can help one in accessing the company's financial and spyware data. It also contains a list of useful video sources to describe design principles and their application in a real environment. Finally, a sample of the weekly educational schedule and assessment strategies is presented.

4.2. THE COMPONENTS OF THE STRATEGIC MANAGEMENT PROCESS

Strategic management system identifies all companies, large or small, and are managed in one way or another, regardless of whether the administrative tasks are officially assigned to an employee or not. Even for companies with a traditional work ethic, long-term business strategy can improve business success.

Figure 4.2. The components of the strategic management process.

Strategic management is the implementation of a large, objective scale using three key elements: environmental research, design and implementation strategies, and evaluation methods.

4.2.1. Environmental Scanning

The first step in the strategic management process is an environmental survey, sometimes referred to simply as "digitization." Basically, this is the process of seeing and quickly managing everything that can impact your business and operations. Internal and external reasons can affect your business. The content may be obvious at first, as managers are often accustomed to what is happening in the company. For example, if a company is facing a shortage of employees, this is a problem that management must address. Other types of content include sales, performance levels, and low profits. External factors may require additional effort to locate and control. Smart managers try to store industry news and current data. These factors could predict a change that will affect the company soon.

Another external factors that need to be investigated include economic data, target markets, and the company's competitors. Each of these factors (internal and external) can be part of a complete SWOT study. It is a strategic analysis of the company's strengths, weaknesses, opportunities and threats. SWOT research provides a comprehensive overview of a company's position in industry and the economy, helping to identify steps that can be taken to enhance and enhance its financial position.

4.2.2. Strategic Planning and Implementation

Environmental research provides a wealth of data. Strategic managers use this information and data to develop strategies that can be applied across industries. Therefore, strategic managers develop thinking strategies to use the strengths and abilities identified in the SWOT study. of course, the chosen strategy can avoid or minimize the risk of injury and threats to the company. After the company accepted the implementation of the strategy presented by the manager, the strategic manager developed an effective framework to implement this strategy. Moreover, each job or step in the program is assigned to a specific employee or department. These employees are responsible for achieving certain goals in order to track the growth of the company to a greater goal.

4.2.3. Strategy Evaluation

Implementing a sensible strategy is not sufficient by itself to satisfy goals. Once the company's employees are completing the planned actions, the corporate must also periodically assess the results of these actions. As a part of their process, strategic managers identify relevant metrics which are carefully monitored and assessed to form sure the corporate is on target to satisfy its goals. Therefore usually, the evaluation phase will begin specific, regular reporting periods where managers and team leaders measure progress. This type of scheduled approach helps to form sure nothing falls through the cracks or gets overlooked. Moreover, the strategy evaluation process is crucial in strategic management. This is often how managers and businesses learn what is working and what still must be adjusted to realize the simplest possible results.

4.2.4. Ongoing Communication

Each of the three components of strategic management requires excellent, consistent communication to form sure the company's objectives are met. All the stakeholders during a business must communicate well with one another. Ideally, this communication should offer each party the chance for input. This includes not only a business's employees but also relevant external stakeholders also. Vendors, industry leaders, customers and even legislators may have an impression on the planned strategy. If that is the case, their input should be considered. At a minimum, the corporate should communicate its plans through the acceptable corporate channels.

Today's technology makes it much easier for busy companies to stay external stakeholders informed. Blogs, email newsletters and social media mean it is easier than ever before to speak clearly with key constituencies.

4.3. ESSENTIAL TASKS OF STRATEGIC MANAGEMENT

Companies need to think strategically so as to survive and thrive during this highly competitive world. There are five essential tasks of strategic management. They include developing a strategic vision and mission, setting objectives, crafting tactics to realize those objectives, implementing and executing the tactics, and evaluating and measuring performance.

4.3.1. Developing a Strategic Vision and Mission

The first step, developing a strategic vision and mission, involves putting management's long-term view of where the corporate goes on paper and making that known to the workers of the corporate. This step is formed from both the vision and therefore the mission. Moreover, the mission is employed to define why the organization exists. Often companies within an equivalent industry will have similar mission statements. The vision is management's view of where the corporate goes. This task is typically the responsibility of the CEO. Executive leadership should not be pulling the mission and vision from nothingness. Hence, there is a mission and a vision present for all companies.

Figure 4.3. Essential tasks of strategic management.

Sometimes the vision is present only as ideas within the heads of top management. Management must determine what it is and communicate that to the workers. Therefore, a transparent and concise mission and vision will help the corporate work together for an equivalent purpose. It is hence utilized in the opposite steps of strategic management.

4.3.2. Setting Objectives

The second step, setting objectives, takes the strategic vision and creates specific goals which will occur to accomplish what is laid out at the vision statement. Therefore, goals should be feasible but also not easily attainable. These sorts of goals are called "stretch" goals because they force the corporate to travel as far as they will to achieve the goal. If goals are set too low, complacency will occur. Moreover, e stablishing these goals removes confusion employees may wear what should be accomplished. Managers can set both financial objectives and strategic objectives. Financial objectives are people who state what management wants to realize in "dollars and cents."

Strategic objectives are goals that strive to extend competitive position, gain market share, or to develop a competitive advantage. Strategic objectives have the facility to motivate and prompt action where financial objectives are seen more as a constraint. Managers should therefore define both goals but consider strategic objectives will cause better results. Both long-term and short-term objectives should be set. Moreover, short-term objectives will motivate present performance while long-term objectives will put the corporate within the proper position to realize what is outlined within the vision down the road. Priority should be placed on long-term objectives over the short-term. Companies that stress short-term objectives find yourself in business short-term.

4.3.3. Crafting Tactics to realize Organizational Objectives

The third step, crafting tactics to realize organizational objectives, is where management defines the way to achieve the defined objectives. During this step, management decides how best to reply to changes within the environment, the way to rise above the competition, and the way to maneuver towards the company vision and strategic objectives.

4.3.4. Implementing and Executing the Tactics

The fourth step, implementing and executing the tactics, includes determining what company resources should be allocated to every activity, establishing policies, motivating employees, providing the resources necessary to realize objectives, and inspiring endless improvement culture. Tactics must be tailored to organizational capabilities and culture for them to figure efficiently. Moreover, change will presumably be needed, but the quantity of change varies counting on how new the tactics are. This usually involves revising policies or resource allocation, moving people around, retraining and retooling, or making changes to reward systems. Hence, each manager must check out the tactics and their department to work out how best to implement each tactic correctly.

4.3.5. Evaluating and Measuring Performance

The fifth step, evaluating and measuring performance, is how management determines whether the tactics were implemented effectively to realize organizational objectives and suits the strategic vision. If performance is not to expectations, corrective action must be taken. Performance is often measured by various methods like financial data, customer satisfaction, quality reports, employee satisfaction, and capital utilization. Each of those sorts of measures should be wont to analyze tactics to urge a full picture of a tactic's success.

4.4. PRINCIPLES TO SUCCESSFUL STRATEGY MANAGEMENT

It is a well-documented incontrovertible fact that 80% of strategic plans fail to understand their value. If you have invested in developing a technique and wondered why it is not producing the results you expected, you would possibly consider a number of these guiding principles to place you back on target. Moreover, undoubtedly, developing and executing strategy (Strategy

Management) can sometimes be quite frustrating, where obstacles seem to constantly get within the way of success. However, over the 20+ years as strategy consultants working with middle market privately held and family businesses, people experienced a good range of situations, from companies that have not yet developed a particular strategy, to people who have attempted it on their own, to people who have retained an outside advisor, all experiencing difficulties in achieving the objectives they began to realize.

As a result, it has been understood that the worth of some guiding principles leadership teams can enjoy when planning and implementing their strategic management process. Hence, bearing this in mind, there are six guiding principles that ought to assist you take advantage of strategy management.

Figure 4.4. Principles to successful strategy management.

4.4.1. Focus on the Foremost Important.

Setting priorities is a crucial step in achieving the strategic objectives because as Stephen Covey of Franklin Covey stated, "the whirlwind of the day-to-day will consume everyone's time. Choose carefully what you think must get done over and above an employee's "day job." More will get through with less stress.

4.4.2. Leverage Strengths

Don't attempt to change the organization into something it is not. Companies got to specialize in, and leverage, existing competitive advantages. Understand your core businesses, what's driving sales and profits, and leverage everything a person gain from it. Balance that with innovation which will reinforce the core and expand the business base to new business opportunities that leverage strengths and capabilities.

4.4.3. Communicate

Be sure everyone in your organization understands what an individual is doing. Make your employees feel they are a neighborhood of the method and therefore the success. Do not let rumors derail the efforts you are making. Keep your employees informed and invite comments and questions. Give your employees reasons to believe that the strategic plan will benefit everyone.

Raise the energy state create a pace of reach create momentum and enthusiasm. Schedule progress meetings on a daily (monthly) basis. Make strategy management a priority and keep everyone interested and committed. The people will become energized by their contribution to the method.

4.4.4. Remain Flexible

Strategy management may be a dynamic and continuous process. Be flexible. Be prepared to form adjustments as competitors, customers and economic market conditions change. Conducting regular team strategy meetings will highlight progress made and action needed on key issues and opportunities.

4.4.5. Invest in Outside.

Help Consider retaining a professional outside resource to assist facilitate the method. This approach allows the Owner/CEO to participate within the process and provides an objective third party which will challenge assumptions and ask the tough questions. As companies grow, the necessity for broader experience becomes greater. an outdoor resource can provide credibility and invite creativity. Moreover, without a question, strategy management is an important part of every successful company. It provides the roadmap for profitable growth. Success ultimately depends upon senior management's ability to supply the type of leadership to both:

- develop a well thought out, fact-based strategy, and
- to foster a piece climate to enable the strategy's execution. This creates a way of ownership and commitment which will ensure its success.

4.5. STRATEGIC OBJECTIVES AND LEVELS OF STRATEGY

Once a strategic analysis has been completed, the subsequent step within the strategy process is to determine strategic objectives. At now, the manager has decided why the corporation exists and the way it will attempt to fulfill its mission. Moreover, strategic analysis has provided information about customer preferences, competitors, and therefore the firm's resources and capabilities. Now it is time to start out planning for fulfillment.

4.5.1. Strategic Objectives

Strategic objectives are the big-picture goals for the corporate: they describe what the company will undertake to satisfy its mission. Strategic objectives are usually some kinds of performance goal – for example, to launch a replacement product, increase profitability, or grow market share for the company's product. Top executives then decide annually what entertainment products the corporate will offer. Because Disney may be a large corporation (more thereon shortly), it is a spread of resources available to make entertainment products to supply. For instance, they will plan to release three movies this year, also to build a replacement amusement park and make five new shows for his or her television network. the strategic objectives at Disney are far more complex than this, because a number of these choices involve long-term efforts (they cannot build a topic park in one year).

Figure 4.5. Strategic objectives and level of strategy.

4.5.2. Levels of Strategies

Once a firm has set its objectives, it must then address the question of how it will achieve them. A business-level strategy is that the framework a firm uses to arrange its activities, and it is developed by the firm's top managers.

Moreover, samples of business-level strategies include cost leadership and differentiation. Hence, these strategies are pursued by businesses with one product or a variety of products. For example, imagine that you simply own a cafe.

An individual is not Starbucks – you are an area shop in your neighborhood, and you run it yourself. Therefore, if one has got employees, but is the manager, owner, and all-around administrator. Hence, while developing your vision and mission statements, one has already made some basic decisions about how your shop will operate. for instance, one has chosen to either offer quick, inexpensive coffee (cost leadership) or a full-service coffee experience (differentiation). Moreover, that call impacts whether you select premium or discount suppliers, how your shop is decorated, and the way many employees you've got to supply attention (service) to your customers. A business-level strategy guides a corporation in how they approach the activities within the value chain.

Operations, for instance, would specialize in efficiency for a price leader and specialize in adding value for a differentiator. When you develop strategic objectives for your shop, you will decide whether or not you would like to undertake to draw in more customers (grow), maintain your business at its current level, or shrink your business (perhaps you are feeling you don't have enough time to spend together with your family). If a person opt that the objective is to grow, for instance, he or she ought to set a selected target, say, to grow revenue by 10%. Once he or she set that specific objective, one will exhibit exactly what business-level actions one must fancy to reach that focus on. Even if a business is far larger than an area cafe, the strategic objectives pursued by these larger companies are not significantly different in concept.

Nevertheless, large companies like Nike or Apple, which have many various business units, develop strategies at several levels. Each individual business unit (say Nike Basketball) will have a manager who decides the objectives for that unit, even as within the cafe example. However, the corporate as an entire will have a chief military officer (the top manager for the company) who develops strategy for the whole corporation. Moreover, corporate strategy is the broadest level of strategy, and cares about decisions about growing, maintaining, or shrinking very large companies. At this level, business-level strategy activities, like an ad campaign to draw in new customers for one line, are not getting to be enough to significantly impact the corporate as an entire.

On the other hand, the corporate CEO essentially manages a gaggle of companies (unless the firm operates together with a business unit) and develops strategies to make success for the general group. consider the group of companies as an investment portfolio: investors attempt to have a various set of investments to spread risk and maximize the performance of the general portfolio. On any given day, an investment that is not doing so well should be offset by one that is doing well. Moreover, a corporate strategy tries to realize an equivalent thing, and CEOs need to weigh the pros and cons of every business unit and the way it is contributing to the success of the general corporation. For instance, a corporation that has business units that had best within the winter (ski resorts) will attempt to even have business units which will perform within the summer (swimming pools) to scale back the danger of getting periods of low revenue. Hence, one tool that corporate strategists use to know how each of their businesses contribute to the corporation as an entire is that the BCG Matrix. The BCG Matrix gives managers a fast picture of which business units do well, and which are not.

Nevertheless, the tool has recommendations for businesses in each quadrant – for example, a business within the dog quadrant should be sold or closed. Cash cows provide income to the corporation, and stars provide growth. A CEO is usually trying to balance the group of business units throughout the quadrants to maximize overall corporate performance. Therefore, it must be noted that the BCG Matrix is not applicable for firm's that operate in one business unit. In order to realize the size of growth necessary to satisfy corporate strategic objectives, a CEO must find ways to develop entirely new business units or reach brand-new markets. For instance, for Walmart to grow their 2017 revenue by 5%, they might have to add $25 billion in new revenue. That is more revenue than opening some new stores could generate.

4.5.3. The Grand Strategy

At all three levels, companies choose a grand strategy in response to the primary question they ought to ask themselves: does the firm want to grow, strive for stability, or take a defensive position within the marketplace? Often, the selection of a grand strategy is predicated on conditions within the business environment because firms generally want to grow unless something (like a recession) makes that difficult. Note that a grand strategy and a company strategy can overlap significantly.

However, a growth strategy involves developing plans to extend the dimensions of the firm in terms of revenue, market share, or geographic reach (often a mixture of those, as they will overlap significantly). For example, Walmart is implementing a growth strategy with the acquisitions discussed within the corporate strategy section. A stability strategy may be a strategy for a corporation to take care of its current income, market share, or geographic reach. Moreover, a firm usually works to take care of a stable position when the choice is to lose ground in one among those categories, for instance due to competition or economic factors. Hence, in today's business environment, publicly held firms rarely aim solely to take care of the established order, because shareholders and therefore the stock exchange reward firm growth.

On the other hand, firms pursue defensive strategies within the face of challenges. a corporation that is struggling may plan to shrink its operations to scale back costs to survive, for instance. A corporation facing strong new competition may need to radically rethink its product offerings or pricing to not lose an excessive amount of market share to the newcomer. Therefore, a technological innovation may make a company's products obsolete (or a minimum of less attractive), forcing it to figure to catch up to the new technology. Ford made a defensive decision when it recently decided to prevent selling sedans within the US due to slow sales compared to trucks and SUVs.

4.5.4. Operationalizing a Grand Strategy

A firm operationalizes its choice of a grand strategy differently at each level of strategy (business, corporate, international). At the business level, a growth strategy means the manager will need to develop ways to grow the business by developing new products or expanding the customer base for existing products, either reception or abroad. Moreover, expanding an organization can take a wider sort of forms. The CEO can develop new businesses, expand to new countries, acquire or merge with competitors, or perform previously outsourced activities.

International expansion is often accomplished by exporting goods to a different country or by acquiring an identical firm in another country to determine the company's presence therein country. Altogether three of those cases, the grand strategy would be growth, and therefore the strategic objectives might be expressed in terms of revenue growth, profit growth, market share growth, or maybe share price growth. (Figure) outlines how a grand strategy often wants to develop specific company actions.

- What is the difference between strategic objectives and a strategy?
- Describe the three levels of strategy and what a manager developing strategy at each level cares with.
- What is a grand strategy and the way does it relate to strategic objectives and therefore the three levels of strategy?
- What are the three grand strategies, and why would firms pursue each of them?
- What are strategic objectives, levels of strategy, and a grand strategy? How are they related?
  - Strategic objectives are the big-picture goals for the corporate: what the company will do to undertake to satisfy its mission.
  - These goals are broad and are developed to support top management's choice of a generic competitive strategy and grand strategy for the firm. For instance, cost-leadership and growth competitive and grand strategies would require managers to develop objectives for growing the firm during a low-cost way.
  - Business-level strategy cares about positioning one company or business unit that focuses on one product or line. The first business-level strategies are cost leadership and differentiation, also as focus, which is combined with one among the opposite two strategies (focus-cost leadership, focus-differentiation).
  - Corporate-level strategy cares with the management and direction of multi-business corporations. These large firms make decisions about what businesses and industries to work in order that they can improve their overall performance and reduce the danger they might face if all their operations were concentrated during a single business or industry.
  - Corporate CEOs use the BCG Matrix to gauge their portfolio of companies and use corporate actions like acquisitions to form significant changes to their companies. International strategies are often combined with either of the previous two strategies to include international operations into a business or corporation. International strategy answers

questions of what country or countries to work in and the way to achieve success in foreign operations. Grand strategies outline an approach to firm growth.

The three grand strategies are growth, stability, and defensive, and a firm chooses one among these approaches additionally to their choice of business-level, corporate, and/or international strategies. The selection of grand strategy is usually dictated by conditions within the business environment like recessions or competitor activities.

4.6. HOSPITALITY AND TOURISM STRATEGIC PLANNING

A strategy is an action plan that a given organization comes up with to realize its business goals. This is often based on how different engagements are going to be linked together to realize success (Berry, 1995, p. 7).

On the other hand, a honest strategy helps the organization in achieving unrivalled growth that enhances sales, profitability and market leadership within the end of the day. Strategic management is meant to help in deciding which will enhance the achievement of future objectives (Heskett, 1986, p.12). In other words, strategic management is the only way that an organization's goals, missions and visions are specified mostly in programs and projects in order that the set objectives are fully achieved. Moreover, within the course of embarking on strategic management, resources are alleged to be allocated to assist in achieving the set programs and projects. After programs and projects are set, there should be a scoreboard which will be utilized in evaluating if the anticipated targets are achieved (Kotter, 1982, p.9).

Strategic management more therefore provides the general direction that a corporation should absorb a bid to make sure that there's unrivaled success. Those within the management level are alleged to come up with good tactics which will enhance the achievement of the set goals. This is necessary because a correct framework on how this may be achieved will ultimately be enhanced (Gladwell, 2000, p. 6). It will only be consistent when those actions enter line with what the management expects to urge as a result of this process. For a strategic management to be effective and achieve what it had been set to, it must follow a process which will enhance this. This is often because it's an ongoing process which will have an extended term effect on the business and industry as an entire.

Figure 4.6. Hospitality and tourism strategic planning.

The steps in strategic planning are as follows.

4.6.1. Develop Strategy, Vision, and Mission

This is the primary step during a strategic management process as there's got to develop a technique, vision and mission which will give direction on how the entire process is going to be undertaken. This may ultimately guide and provide a roadmap that creates its execution easily (Christensen, 1997, p. 13). The strategy should be in line with the short term and future objectives. The mission will explain the role that the industry is probably going to play within the market. With this in mind it will be easy to realize the set objectives.

4.6.2. Analyze the Interior and External Environment

It is important to try to do an analysis of the environment under which the strategy is going to be administered (Hamel, 2002, p. 5). Therefore. the environment plays a crucial role in determining if the strategies are going to be successful as planned. It is vital that both the interior and external environments be checked out to understand where to enhance on and achieve the set objectives. During this case, there is a need to do a self-evaluation in order to see that the interior environment is well understood or not (Gladwell, 2000, p. 4). In order to cater to the external environment, a competitor analysis should be done in order to understand what to expect. This is often also necessary in telling whether the strategies are going to be successful or not.

4.6.3. Refine Options and Choices

One should have options and within the process make choices in arising with good strategies in strategic management (Traverso, 2000, p. 15). This is often necessary in coming up with the simplest alternatives which will add variety and code to the industry. However, in refining options, it will be easy to skim through the varied weaknesses that every option has. After this has been done it will be easy to settle on the simplest amongst the varied options.

4.6.4. Plan

A plan is important in giving a road map that the entire process will undertake. This may provide details on how the set objectives are going to be achieved (Cokins, 2004, p. 17). The plan will keep the entire process of strategic management organized. This may help in knowing what to try to do next because the plan stipulates. within the process they are going to be before time in implementing strategies. This is often necessary for the fulfillment of the project at the end of the day.

4.6.5. Implement

After all this has been done it is necessary to implement what has begun within the plan. It is only through implementation that one can tell if the entire process is going to be successful. Implementation is going to be done systematically in order that all the steps are administered as this is often necessary for fulfillment (Lynch, 2009, p. 17). Moreover, a proper implementation will lay an honest ground for strategic management. Additionally, it is only through implementation that the management is going to be ready to know the hurdles that need to be overcome in completing strategies.

4.6.6. Evaluate

Evaluation helps in measuring the effectiveness of the strategy that has been put in situ. This is often in terms of telling if what had been begun was ultimately achieved (Morgen, 2003, p. 17). Proper evaluation is often done by watching the strengths, weaknesses, opportunities and threats in situ. On the opposite hand, evaluations are often assessed on how suitable the strategy was in achieving its objectives. Feasibility is important to inform if the resources were enough in successful implementation of the strategy.

4.7. IMPORTANCE OF STRATEGIC MANAGEMENT DURING A HOSPITALITY BUSINESS

The hospitality industry must have a strategic management process. Strategic planning lies within the value of designing before time in order that the industry can perform well. This is often because if they fail to plan, they are going to fail (Heskett, 1986, p. 9). The industry must have a strategic management process for unity, direction and a way of identity in achieving business goals. Moreover, strategic management will help the industry to realize their organizational goals in an efficient and effective manner. This is because there is a mission and vision that require to be attained in both the short term and future. The industry is ultimately continuing to grow and wishes direction.

Figure 4.7. Importance of strategic management.

With strategic management in situ the industry is going to be ready to come up with more complex management structures that are necessary to facilitate its growth and development (Cokins, 2004, p. 13). this will be effectively through with a well laid out plan that features an outlook. For the industry to speak in one voice it must have an honest strategic management outlook (Christensen, 1997, p. 9).

This is often necessary in ensuring that the environment that they operate in is favorable to sustain the companies. The industry is interlinked in some ways and wishes to speak in one voice during a bid to strengthen their position within the market. It is also necessary because it helps in deciding as there will be laid down avenues for efficiency (Kotter, 1982, p. 7). Moreover, decisions that are made will help in reassessment and implementation of changes which will positively transform the industry during a broad perspective. additionally, the management from various players within the industry will have a chance of accessing their performance.

4.7.1. Current Strategic Effort

In order to become a world class leader within the provision of food and support services the corporate has some strategic outlook. Currently the firm features a strategic effort in situ for sustainability. Its strategic effort is in setting high standards for responsible business practices (Compass group, 2010, p. 12). This is in terms of business conduct that involves safety, health and environmental practices. The corporation plans to be a coffee cost and efficient service provider to offer it an honest market outlook. Additionally, the corporation seeks to develop the prevailing expertise and strength in contract food services.

4.7.2. Environmental Elements

The company operates in many markets and as a matter of fact is exposed to different business environments. Internally, it is put in the proper management practices to make sure that its employees play a crucial role in delivering business goals and objectives (Compass group, 2010, p. 5). That is why they value the range and individuality that every employee brings within the company. More so, it operates in an environment of trust which has cultivated good management practices within the group.

Nevertheless, through proper management of assets entrusted to them by investors, the corporate has ensured that it generates active rates of returns. By doing this, the corporation has always received support from investors altogether for its expansion strategies. The external environment has played a task in shaping how the corporate carries out its activities (Compass group, 2010, p. 6). Therefore, this environment is vital in determining the general success of its global ventures. Moreover, compass has respected the legal frameworks (in countries it operates) by abiding by them. Although there are some legal constraints, it is soberly engaged governments in ensuring that there is a level playing field. There are some economic factors that the corporate has faced in executing its operations (Compass group, 2010, p. 6). Hence, different countries have different rates of economic process and therefore the company has factored these in its expansion plans. Exchange rates have varied; this has affected its revenues during a broad way. The corporate features a good conversion mechanism in situ for continued sustainability.

4.8. THE CURRENT EFFECTS OF THE STRATEGIC INVESTMENT POLICY IN DOMESTIC ECONOMIES AND TOURISM POLICY

Tourism and travel are predicted this year (2011) to form 6.5 billion denars or € 105 million, as an immediate contribution to the country's economy. This amount is merely 1.4% of total gross domestic product (GDP), consistent with the report of the planet Council for travel and tourism. The Council for Tourism may be a forum of business leaders from the tourism and transport industry during which executives of 100 largest global companies from these industries are directly involved. Moreover, the council monitors the info in terms of tourism for 20 years together of the world's largest industries, which on the average contributes 9.2% to world GDP and employs 260 million people worldwide. According to these data, Macedonia is on the 168th place within the world, consistent with what proportion tourism contributes to the country's GDP. All our neighboring countries are before Macedonia.

Nevertheless, in Bosnia tourism contributes 2.31% to GDP, in Bulgaria 4%, 8% in Montenegro, and while in Croatia even 11.64%. World average, in terms of what proportion is that the direct contribution of tourism to the GDP of nations within the world, is 5.12% and that we are far below this average. The situation is analogous when it involves tourism contribution to the entire employment within the country, also as investment in tourism. While in Croatia tourism directly employs 12.7% of total labor, 8% in Greece, Montenegro 6.9%, Bulgaria, 4.9% and a couple of.1% in Bosnia, in Macedonia just 1.3% of the entire number of employees are directly involved in tourism in terms of investment, the country is on the 156th place within the world where we are again last within the region.

However, the council of Travel and Tourism predicts that this year in Macedonia is going to be invested only $ 50 million, versus $140 million in Bosnia, $270 million in Montenegro, $620 million in Bulgaria, Croatia $1.41 billion and seven.41 billion dollars even in Greece and even greater amounts in Turkey and Italy, which are incomparable with our country. The Council has somewhat better forecasts regarding the event of Macedonian tourism within the next ten-year period till (2021). Moreover, the contribution of tourism to national GDP is considered that it'll grow within the next 10 years from 1.4% to 4.9% which would bring revenue within the amount of 10.5 billion consistent with today's prices. The Council also considers that investment in tourism will rise to the extent of 4.7 billion and would amount to 2.8% of total investment, compared to 2.3% today.

On the opposite hand, it is not expected a big increase in persons employed in tourism with their number expected to rise to 10,000, compared to eight,000 people, who are employed today. The country is often said: "The Republic of Macedonia remains relatively on a coffee level in attracting and using foreign direct investment, compared to other countries within the region. However, in order to extend the influx is important to make a stable economic environment and strengthen the private sector." Therefore, Macedonia should provide input to foreign direct investment because it ensures intensification of economic development, job creation and poverty reduction.

4.9. OPERATION OF THE MANAGEMENT IN THE BENEFIT OF TOURISM ENTERPRISE

Within the current complex situation of the economy in Macedonia, tourism is criticized for its lack of organization and performance. Hard-earned tourism penny by catering and tourism workers, is hardly used for themselves or for upgrading of their services and expanding them add the non-use of promotional activities, both domestically and abroad. The approved money for propaganda and development of tourism is difficult to understand, for the straightforward reason that tourism in our country is not sufficiently organized, left to itself.

It is not learnt from the experiences of nations during which tourism is taken seriously, and therefore the state accomplishes an honest influx of foreign currency. With frequent changes in management structures in Macedonian economy, tourism is increasingly losing its value as an economic branch, which is perhaps unique in using only domestic raw materials. Hence, the most strategic activities in tourism; the entire process of working to use domestic employees, homemade food, including an increasing number of hotel-and catering organizations, small private businesses like catering and other entertainment facilities, an outsized number of sales and manufacturing firms for production and sale of souvenirs and native handicrafts.

Strategic management in tourism must also include the extra activities, which suggests a development of tourism like transport, postal services, and other services and manufacturing facilities. The Republic of Macedonia has many possibilities to supply a hotel accommodation for transit tourists from central and southern Europe, as Skopje and Macedonia are the foremost convenient places to rest after having spent thousands of kilometers. For this purpose, with good strategic management, a separate tourism brochure should be made for transit tourism in Macedonia, which should be promoted to our presence abroad – at fairs, congresses, tourism events, etc. The geographical position of Macedonia is often used as a big competitive advantage for these motorized tourisms, too. The territory of Macedonia is the best suited place for the remainder of the thanks to Greece, but little is invested and managed all.

Macedonia as a state, although extended on a little piece of land, has many natural and cultural treasures expecting someone to gauge, to place them in position to become a tourist attraction and take advantage of their visit.

4.10. IMPLEMENTATION AND THE STRATEGIC PROCESS

Implementation is the system-wide action taken by firm members aimed toward accomplishing formulated strategies (Hahn and Powers, 2010). Most people intuitively understand that tons of data are required before any big decision is formed. For example – consumers would not normally buy most the products, say a vacation, a hotel booking or attendance at a festival (all relatively expensive purchases) without investigating the attractions of the destination, hotel or event and typically checking out something about the corporate providing the service before purchasing.

Buying, say, a pencil would not warrant such analysis since it might usually represent a little proportion of the buyers' income. In the same way, a corporation would be risking an excellent deal if it were to pursue a strategic option without first completing an in-depth analysis of its internal and external environments. Put simply, successful strategy selection and implementation relies upon the pre-supposition that the organization has administered a meaningful strategic analysis and is consequently conscious of its internal strengths and weaknesses and its external opportunities and threats. Without being 'armed' with this information, the corporate cannot be certain that the chosen strategy would be the right choice.

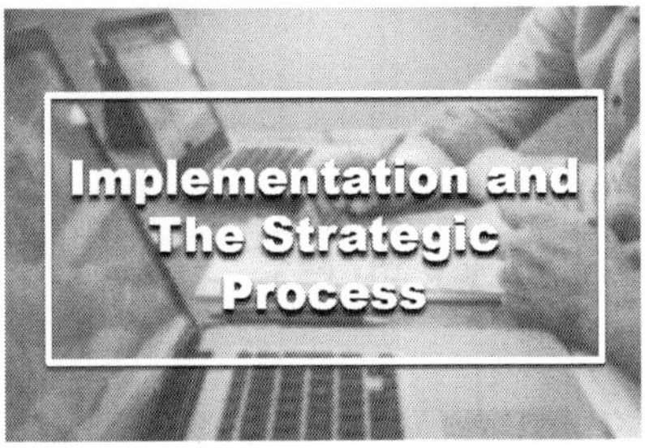

Figure 4.8. Implementation and the strategic process.

It is not surprising perhaps, since it is more geared towards action and fewer conceptual in nature, that the implementation phase of strategy has received relatively less academic attention than the analysis and formulation phases (Kaplan and Norton, 2001:1; Evans, 2005; Hahn and Powers, 2010). It's also been argued that the so-called 'quality gurus,' like Tom Peters (Peters and Waterman, 2004), largely abdicated responsibility for delineating patterns of implementation which are coherent (Morris and Haigh, 1996). Other authors make important contributions by pointing to the difficulties inherent within the implementation phase (Epstein and Manzoni, 1998) or by calling for a re-assessment of the difficulties inherent within the process (Lorange, 1998).

However, the very fact that successful implementation is vital to a successful strategy has long been recognized. Several authors (Ghoshal and Bartlett, 1999; Miller, 2001) point to the shortage of successful strategic implementation as a problem affecting many businesses. In their study Bartlett and Ghoshal (1987: 12) means that the problems involved did not relate to a poor understanding of environmental factors or an inappropriately strategic intent. Instead, the authors report that 'without exception [the organizations within the study] knew what that they had to do; their difficulties lay in the way to achieve the required changes.' Miller et al. ask the 'implementation gap' in arguing that 'organizations are slower to vary and harder and more expensive to develop than strategies are to prepare.' Consequently, given its importance within the successful delivery of outcomes, the strategic implementation phase has begun to attract more attention in recent years (Okumus (2003), Evans (2005) and Hahn and Powers (2010)).

4.11. ASPECTS OF STRATEGIC IMPLEMENTATION

In order to successfully put into practice (implement) a technique, a corporation will have to consider three aspects: resources, configuration, and change.

4.11.1. Resources

How should the strategy be resourced? This relates to the way during which the organization will obtain the requisite finance, human resources (usually within the sort of appropriately skilled employees), the physical resources like equipment and buildings, and intellectual or 'intangible' resources. All parts of the organization got to be aligned in order that all parts are working towards a standard vision, aim and objectives. In doing so, detailed decisions about obtaining and utilizing resources need to be made so as to make sure that resources are used effectively, and waste and duplication are avoided.

4.11.2. Configuration

How should the culture and structure of the organization be configured to 'fit' the proposed strategy? Within the point we stressed that each one part of the organization got to be aligned in order that the strategy can successfully achieve its vision, aim and objectives. However, since all parts of the organization got to contribute to its successful implementation, it's necessary that the structure of the organization and therefore the organizational culture are fit for purpose in order that there are not any impediments to success. This is often not always the case and in many instances the structure must be amended and therefore the culture changed in order that there is a so-called 'strategic fit.'

4.11.3. Change

How should the changes arising from the strategy be managed and led? Implementing strategy invariably involves changes to aspects of the organization and the way it positions itself. To successfully implement change, managers got to consider what sort of change is envisaged and the way it'd be managed and led. Thus, strategic implementation is involved making detailed decisions regarding three key aspects of strategy relating to: resources; configuration and alter. When these decisions are made it's vital (since all parts of the organization got to be aligned for successful implementation), that an extra issue is taken into account.

This relates to the dissemination and coordination of the strategy. Some research suggests that a lot of strategies are well perceived but poorly executed and Jones (2008) in his book which focuses on communicating strategy reports on research which suggests but 10% of employees understand their firms' strategy.

Thus, crucial questions arise as to:

- How should the strategy be spread throughout the organization (or 'cascaded' because it is usually termed)?
- How should the implementation be coordinated effectively, during a practical way?

This involves:

- communicating the strategy effectively.
- coordinating the implementation of the strategy consistently and effectively;
- and measuring the success of the strategy in order that modifications are often made if necessary.

Thus, the strategy must be communicated and disseminated. As a part of this process an in-depth strategy could also be developed for every part of the organization – like marketing, finance, personnel and strategic business units – since successful implementation normally involves all elements being aligned with specific objectives. Effectively organizing strategic implementation may be a major managerial task and during this chapter we will consider one widely adopted means of managing this process – The Balanced Scorecard.

4.12. CONCLUSION

This chapter throws light on the promotion of tourism, how through the framework of planning and after the proper analysis of all the needs in the industry, the organization reaches to its goals. This chapter shows strategies and process for the same by describing the components of strategic management process to the environmental scanning process. It also elaborates the strategic planning and implementation and the strategy evaluation. It sheds light on the importance of ongoing communication and how the establishment of strategic vision and mission helps the industry to grow.

As the chapter proceeds, there is portrayal of how the organization survives in the world which is so competitive by giving a glimpse upon the essential tasks of strategic management and the crafting tactics to realize the organizational objectives and to implement and execute these tactics. It also deals with the importance of strategic management in a hospitality business and the recent effects of the strategic investment policy in domestic economies and tourism policy. The strategic process, its implementation and the different aspects of such strategic implementation like resources, change and configuration.

REFERENCES

1. Albu-strategymanagement.com. (2021). *6 Guiding Principles to Successful Strategy Management* – Albu Strategy Management. [online] Available at: https://albu-strategymanagement.com/2015/07/6-guiding-principles-successful-strategy-management/ (accessed 8 August 2021).
2. Annie Sisk, (2018). *The Three Components of the Strategic Management Process*. [online] Bizfluent. Available at: https://bizfluent.com/info-8238476-three-components-strategic-management-process.html (accessed 8 August 2021).
3. Eric Vanderburg, (2004). *5 Essential Tasks of Strategic Management.* [online] Medium. Available at: https://medium.com/@evanderburg/5-essential-tasks-of-strategic-management-57d7bc097cc7 (accessed 8 August 2021).
4. Evans, N., 2019. Strategic Management for Tourism, Hospitality and Events. [online] Available at: <https://www.taylorfrancis.com/books/mono/10.4324/9780429437601/strategic-management-tourism-hospitality-events-nigel-evans> [Accessed 13 August 2021].
5. IvyPanda, (2019). *Hospitality and Tourism Strategic Planning* – 3191 Words | Essay Example. [online] Free Essays. Available at: https://ivypanda.com/essays/hospitality-and-tourism-strategic-planning/ (accessed 8 August 2021).
6. Kirovska, Z., (2011). *Strategic Management Within the Tourism and the World Globalization.* [online] Ideas.repec.org. Available at: https://ideas.repec.org/a/ris/utmsje/0028.html#:~:text=Strategic%20management%20is%20a%20proactive,is%20affected%20by%20tourism%20development. (accessed 8 August 2021).
7. Moutinho, L., & Vargas-Sánchez, A., (2018). *Strategic Management in Tourism.* 3rd ed. CABI.
8. Opentextbc.ca. n.d. *Strategic Objectives and Levels of Strategy*. [online] Available at: https://opentextbc.ca/principlesofmanagementopenstax/chapter/strategic-objectives-and-levels-of-strategy/ (accessed 8 August 2021).

Chapter 5

Tourism Demand Forecasting: Parameters and Techniques

CONTENTS

This chapter of tourism demand forecasting explains the significance and use of various parameters and techniques that have been used in tourism industry. It also explains the importance of the tourism and transportation demand forecasting. This chapter also provides highlights on the various types of forecasting methods. This chapter provides an overview of the four-step travel modeling process. The chapter also addresses an integrative approach to tourism forecasting.

5.1. INTRODUCTION

With an ever-growing rate, the tourism industry seems to have grown in the last few decades and for a country's economy it is has become an important sector. In several countries, the contribution of the tourism and passenger transportation industry is quite huge towards the economic growth rate and various factors like the foreign investment and job creation can be impacted immensely by it. Towards the end of the twentieth century middle income economies that were emerging saw a large spread in mass tourism. Moreover, as per the UNTWO World Tourism Barometer's report, when compared to 2006, the arrival of international tourists globally saw a six percent growth from January to April in 2007 wherein the strongest growth was seen in the Pacific and Asia (Chu, 2009). Hence, as the economy improved after the crisis of 2009, more and more people planned and traveled to various regions all over the world. The world tourism rebounded to its growth which was for the longer term as a result of this main factor (Croce, 2018).

Nevertheless, the World Travel and Tourism Council estimated in 2014 everyone in eleven jobs was tourism based and about 10% of the gross domestic production (GDP) globally is contributed to by the tourism industry. The latest report of this organization came out in 2018 which showed that of the total employment 9.9% or 313 million jobs and 10.4% of the GDP globally was accounted for by tourism and travel (WTTC, 2018). Moreover, for several regions and countries across the world the economic impact of the tourism and passenger transportation industry is quite vital. To take an example, only 250,000 tourists arrived in Vietnam in 1990 and by 2011 this number had swelled to six million.

In that very year 1.3 million jobs were created by tourism in Vietnam which accounted for five percent of the GDP of Vietnam (Nguyen et al., 2013). In 2006, Egypt was visited by nine million tourists whereas only 2.6 million tourists had visited the place in 1990 (Zaki, 2008). For Malaysia,

5.6% of its GDP was contributed by the tourism industry in 2012 (Arsad and Borhan, 2014).

The very fact that in Catalonia in Spain the tourism and passenger transportation industry contributed to the region's twelve percent of the GDP and fifteen percent of the working population's jobs shows the tourism industry's importance in this region (Claveria and Torra). Moreover, 10% of Singapore's annual $ 87 billion economy is made up by tourism whereas the tourism and passenger transportation industry related jobs are as high as about 150,000. In 2007, the tourists in Singapore were double than its population amounting to eight million tourists (Chu, 2008). In New Zealand during the period from 2000 to 2012 there was a growth of fifty percent in the tourism and passenger transportation industry. 19.1% of the total jobs in New Zealand during 2012 were tourism based and the tourism industry contributed directly or indirectly to the tune of 14.9% towards the GDP (Huang et al.,2014). During the last few decades, the tourism and passenger transportation industry has seen a steady rate of growth barring 2008 when major economic crises hit the nations (Ibrahim and Loganathan, 2010).

On the other hand, a demand management plan that is appropriate is required to be made by the private sector as well as the government agencies as the growth and importance of the tourism industry in the country's economic development and its own growth is only increasing. Both an overestimation as well as underestimation of the volume of future demand owing to miscalculations can turn out to be quite costly.

Figure 5.1. Parameters and techniques play a significant role in tourism demand forecasting.

Facilities shall depreciate fast as a result of stations and facilities getting congested with crowds leading to dissatisfaction amongst the customers and ultimate loss of customers due to underestimation of demand. On the other hand, the cost of the business shall increase unreasonably due to the creation of idle capacity, maintenance of unsold seats that are of high quality and the total overhead costs shall increase as a result of overestimation of the demand (Sharif Azadeh et al., 2013). Moreover, when a strategy is developed for tourism, it is essential to analyze the requirement of transportation as the tourism industry inherently depends on the transportation of passengers. For the planning of transportation facilities, forecasting studies become essential due to the transportation's importance in the tourism industry. However, an operational and development plan that can result in an increase in tourism and passenger transportation industry's profitability necessarily requires an accurate forecast of the demand for tourism (Lai, Yeh and Chen, 2012).

5.2. TOURISM AND TRANSPORTATION DEMAND FORECASTING

The aim of forecasting demand for the tourism and passenger transportation is to find estimations that are nearly accurate of the services or products that may be demanded as when a plan is being made to supply a service or product an estimation is required as to how much of the service or product shall be consumed or purchased. Therefore, requisite facilities for the tourists can be provided with the help of forecasts that have been made for the long-term and short-term as well. Moreover, expensive facilities and infrastructure like road extensions depending upon the hike in demand for tourism for certain attractions can be developed with the help of long-term forecasts.

On the other hand, the use of short-term forecasts is more apt for weekly or daily operations like for instance in the case of cruise tour pricing which fluctuates as per the seasonal requirements of high and low. For instance, a visit may be better scheduled by having a short-term weather forecast that is precise and enable one to plan clothing too accordingly. Similarly requisite infrastructure like transportation terminals and hotels can be planned through a long-term forecast of the precise number of visitors that are likely visit in a year. To make a forecast, historical data is used by almost all the quantitative methods and data from the test markets is used by the others. The demand forecast based results may be utilized for planning the workforce, supply, defining the strategy for marketing and for pricing. However, an accurate forecast of the number of tourists likely to arrive in the region is required and desired by various companies like the tour providers, cruise tour operators, resorts and hotels, airlines, shops and a number of recreational facilities' owners.

Figure 5.2. Tourism and transportation demand forecasting.

The capacity for selling products (e.g., souvenirs, food) and services (e.g., seats in an airplane, rooms in a hotel) may be lost due to forecasts that are imprecise. The success of most of these business enterprises is based on proper management of tourism demand. It is therefore vital to keep in mind that due to the perishable nature of tourism there can be major losses in sales of all the business enterprises if it is not sold at the appropriate time (Cuhadar et al., 2014). The main characteristics with respect to tourism forecast that are considered in various studies include: the amount of money spent on an average by each tourist or the total money that is spent by all the tourists in the entire year in the region of the destination; the number of nights spent by each tourist in the targeted destination and total number of trips made to the tourism destination from the country of origin (Teixeira and Fernandes, 2008).

In the tourism industry, a mandatory factor for the strategic planning is the estimation of the volume of the passenger transportation which is tourism's inseparable part (Postorino and Andreoni, 2006; Tsai et al., 2009; Milenkovic et al., 2013; Manrai et al., 2014). Passenger transportation has time and again shown to have an increasing importance in this domain as well as the tourism sector as can be seen in the literature on tourism.

5.3. FORECASTING METHODS

Several pages of review are in existence as of now with respect to the methods for forecasting the demand in tourism and passenger transportation. In this domain, the studies conducted by Karlaftis (2010), Chan (1979) and Brand (1973) are some of the early reviews. Moreover, in 1995, Witt and Witt (1995) conducted a review on the forecasting of demand in tourism and this is one of the most important ones. They brought forth that as timely and accurate forecasting is extremely important for products and services that are perishable in nature and tourism being a perishable product, the forecasts in this sphere need to be accurate.

Previous empirical studies were evaluated by them where the focus was on the methods that were available around that time and came to the result that in all cases a single method cannot outperform the others. The accuracy of the econometrics, exponential smoothing and auto regression was seen to be higher than the no change model amongst the methods that were studied. Similarly, twenty-two studies were examined by Li et al. in 2005 which dealt with the econometric models pertaining to demand forecasting that was international, and the focus of their study was on advance econometric models. They concluded that for short-term forecasts generally the performance of structural time-series model and time-varying parameter model is good. Also, the econometric model performs far better than the conventional no-change model or the univariate time-series when the forecasts are annual data based.

5.3.1. Time Series Models

Constant time intervals document the ordered sequence of a random variable's values in a time series. Based in the time series a stochastic process is created as a result of this documentation. Some examples of time series are the average daily world prices of crude oil or gold; or the amount of rainfall received in a city daily. In order to create a base that can predict a variable's future values only the historical values of the variable are required in time series. Moreover, through the process of time series forecasting the future values of a variable can be predicted using a forecasting model that is based on the values having been observed earlier. Tourism industry demand forecast widely applies the time series models. Hence, in order to predict the demand for tourism in Singapore several time series models were studied in 2005 by Morzuch and Oh. Nevertheless, the conclusions drawn by the authors were quite useful whereby it suggested that: different models may be chosen by the length of the forecast horizon and the performance statistic (e.g., MAE, MAPE, RMSE) as the best model for them; a model may not continue to be the best one in performance post-sample period even though its performance was the best during the within-sample period and the best model may change as per the duration of the period of prediction. They concluded through their study that there are more reasons for error to occur in the forecast through the structural models and their forecast is not as accurate as the univariate models.

5.3.2. Autoregressive Moving Average (ARMA), Autoregressive Integrated Moving Average (ARIMA), and SARIMA

Two models of moving average and autoregressive models make up the ARMA models which are a kind of stationary stochastic models. Most of the applications and development of ARMA class models are with regard to the book by Box et al. (2015). The notation ARMA is used most of the times to show the ARMA models wherein the aggressive part is depicted by p and the order of moving average part is shown by q. In the ARMA models there is a basic limitation with respect to the data being stationary. Rather than the classic ARMA models, the ARIMA models need to be applied when the data is not stationary.

5.3.3. Regression Models

Based upon independent variables that may be one or more a dependent variable can be calculated through the regression models which determine a function for forecasting (Tukey and Mosteller, 1977). For the independent variable, the term "predictor variable" is used and for dependent variable, the term "response variable." A multiple linear regression was attempted by Varagouli et al. (2005) to fit into Xanthi's (in northern Greece) travel demand. So that the demand for travel could be predicted, the authors tried to propose a model which was based on the variables they had tried to identify as affective. The authors were of the belief that most of the times the multiple-step processes were used to forecast traffic however, traffic forecasting was directly studied by them whereby they examined a functional relationship between roadway characteristics and socioeconomic influences.

5.3.4. Support Vector Machines

About the learning tasks like regression and classification a machine learning method, the SVM was developed by Boser et al. (1992). The basic premise behind SVM is that in a higher-dimensional feature space linear trends can be mapped into by non-linear trends in input and a learning algorithm is used to recognize the complex data's subtle patterns (Vapnik, 2013). Moreover, there are two main categories in SVMs: support vector regression (SVR) and support vector classification. Generalized performance is attempted to be achieved by the minimization of the generalization error bound through SVR which is the focus in this section. It is noteworthy that only a subset of the training data is relied upon by the models produced through SVR.

5.3.5. ANN models

> Tourism related studies for forecasting demand in the late 1990s for the first time used neural networks. For forecasting the demand in tourism, a larger number of researchers are now applying this method which is evident from the high usage of ANNs by a number of publications.

Strong assumptions are not required by the ANN models, they are non-parametric and data-driven and nonlinear data needs can be learnt by them which is unlike the classic statistical methods. Generally, an input layer, an output layer and several hidden layers that are varying comprise a neural network. Input variables like the demographic and economic data are represented by the input layer's nodes. For the internal understanding by the network of the trend in nonlinear data hidden layers are used and the problem's solution is represented by the output layer. Processing units of the network are the neurons which are nodes present in each layer of a network whilst being connected to the adjacent layers' nodes. In order to generate an output a transfer function is applied by each neuron to its input variables' weighted summation. Finally, there are two categories of the Neural Networks: feed–forward networks through which the information flowing in only one direction is considered and the recurrent networks through which the feedback considerations from the ensuing level layer to the previous layer of neurons is considered (Synder and Pattie, 1996).

5.4. OVERVIEW OF THE FOUR-STEP TRAVEL MODELING PROCESS

The "four-step" process which is the conventional sequential process is frequently used for estimating transportation demand and this report presents this method.

Step 1 – Trip Generation;

Step 2 – Trip Distribution;

Step 3 – Mode Choice;

Step 4 – Assignment.

The following paragraphs describe the other components that are usually included in the four-step process.

It is not implied by the four-step process' serial nature that the travelers make sequential decisions instead of simultaneous ones or that the order as is implied by the process is exactly followed by the decisions. For instance, whilst deciding upon the mode of traveling the decision about the destination may be made simultaneously or consequently. At the same time, the four-step process does not imply that independent of the decisions made for other trips, each trip can be planned. For instance, the mode of travel used in a previous trip may influence the decision about the mode to be used for the trip that is being planned.

"Trip" is the unit of travel in the four-step travel models which entails the individual or the vehicle traveling without an intermediate stop to the destination from the point of origin. Trip purpose is used to segment the trips in the four-step model as people behave differently for varying reasons. The kind of information that is required by the model so that it can cater for planning analysis, the characteristics of the region that is being modeled and data that is available to obtain the parameters of the model and the model's inputs decide the definition and number of trip purposes in the model. However, most of the models have a minimum of three trip purposes: home-based non-work, home-based work and non-home-based work. These tree-trip purposes are referred to in this report as the "classic three" purposes.

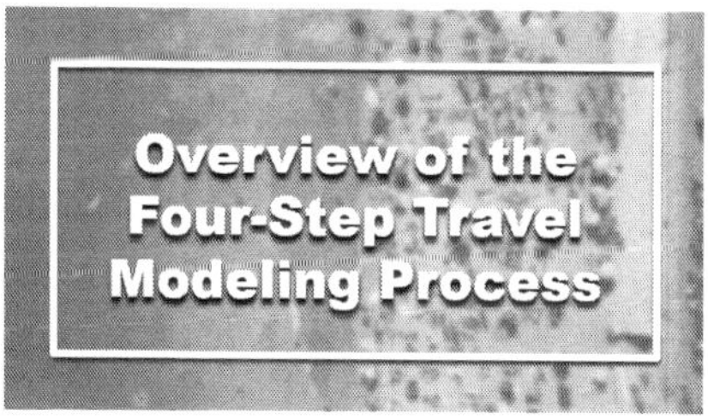

Figure 5.3. Four-step travel modeling process.

Depending upon the amount of activity in the area that is being analyzed, the number of trips of every kind that end or start in each location are estimated through trip generation. In a majority of models trips are collated for a specific geographic unit (e.g., a traffic analysis zone). The flow unit used in the model will have the number of daily trips that have been estimated and this is often one of these: person trips in motorized modes (transit and auto); vehicle trip; person trips through all modes that include both non-motorized (cycling, walking) and motorized modes.

Certain characteristics classify the number of households which are included in the variables like the number of workers, number of persons, availability of vehicles and employment as per type and income level. Traffic zone and purpose-based attractions and the trip productions are the output of trip generation. The question pertaining to the number of travel trips between various units of geography is addressed through trip distribution (e.g., traffic analysis zones). Effectively, the attractions from the trip generation step and the trip productions are linked by it.

The cost involved in traveling between zones (which includes the time fact) as well as the amount of activity involved in trip-making at both the destination and origin zone are the explanatory variables required for trip distribution. By purpose, the trip distribution's outputs are the production-attraction zonal trips. Moreover, the trips that have their origin outside the model area or the geographic region of the model or are destined outside it are estimated by the modes of external travel. Hence, elements of trip distribution and generation are included in these models and the external travel is represented by the trip tables which are the outputs.

In the four-step process, the third step is mode choice. Depending upon the mode of travel the trip distribution step-based trips in the tables output are split in this step. Based on the types of planning analyses that is required and the kinds of options offered for transportation in the geographic region of the model the definitions of the mode vary however generally they can be grouped into non-motorized, transit and automobile modes. Cycling and walking are two of the non-motorized modes that are yet to be included in certain models mainly in the urban areas that are smaller. Service type (express bus, local bus, light rail, heavy rail, commuter rail, etc.) or/and access mode (auto/walk) may be used to define the transit modes. The occupancy level is usually used to define the auto modes (shared ride with two occupants, drive alone, etc.). In order to convert the auto person trips into vehicle trips conducted prior to the assignment automobile occupancy factors are used when there is no separate modeling is done for the auto modes. Person trip tables by purpose, mode and auto vehicle trip tables are included in the mode choice process' outputs.

5.5. MEASURING TOURISM: METHODS, INDICATORS, AND NEEDS

It is widely known and accepted that tourism is a complex phenomenon. Incidentally, many activities, entities, subjects, behaviors or sectors are attempted to be included in the blanket term-tourism whereby by and large they all relate to the movement of people across various countries or places. Therefore, the grouping of such many diverse elements within this concept leads one to question as to whether tourism is so chaotic and varied to require consideration separately as an economic sector or subject (Cooper et al., 2008:5).

The measurement of this phenomenon or the effects thereof form the basis for many actions, policies and strategies at the global or even the local level despite of this complexity about tourism. Moreover, a measurement of the characteristics, a definition and its evolution in time is required as per the cultural tradition whenever an action is wanted or needed with respect to a system or a phenomenon. This is basically, a "scientific" approach (Hepburn & Andersen, 2016). However, a series of difficulties must be considered when tourism is concerned largely due to the reason that complex adaptive systems are dealt with by us. The characteristics of the phenomenon and the systems associated with it that can be recognized relatively with ease give complexity to this rather than the diversity or number of items that are being considered. Basically, components in some large numbers which have different natures are present in it wherein nontrivial relationships exist between the external environment and them and the sensitivity of their evolution (as a group and individually) towards the initial conditions is quite high. "Emergent" phenomena arise from this which are basically configurations or events not being inferred easily from the components individual characteristics and only simulation techniques can be used to foresee them even as they come mostly as a surprise. Furthermore, the system is allowed to adapt itself through the feedback loops or its "memory" as per the feedback or its history (Johnson, 2009). These features are common to all the tourism manifestations and systems as has been shown by many scholars (Faulkner & Russell, 1997; Farrell & Twining Ward, 2004; Baggio, 2008; Baggio & Sainaghi, 2011).

Figure 5.4. Measuring tourism: methods, indicators, and needs.

The inbuilt unpredictability in the system's dynamics is one of the main outcomes of dealing with systems that are complex (Boffetta et al., 2002). Therefore, a common hypothesis that is assumed implicitly even though it is not declared in most of the forecast works is that inertia is exhibited by all systems through which they are driven along an evolutionary path that is stable and temporary. This basically implies that it is impractical to have long term predictions that are correct but the method that have been devised may be used to attempt forecast within the limitation of not taking the forecast for extended periods of time (Somette & Andersen, 2005; Tippett & DelSole, 2009).

5.5.1. Current Methods and Main Issues

With the gaining importance of tourism in the economy it is but natural that in terms of the demand and supply that reasons and an economic terminology be adopted. All the people who travel to a common place comprise the demand (destination and tourists). Four elements can be considered in order to measure this namely: money (receipts and expenditure); space (lengths of trips and distances); time (travel durations and stays) and people (tourists) (Song et al., 2010). Moreover, money and people are the two main measurements considered. Hence, collection methods and sources for demand data differ immensely across the world even though a lot of effort has been put into this aspect by many organizations nationally and internationally like by recommendations made by the statistical division of the UN, the European statistical office and the UNWTO (UNStats, 2008a; EUROSTAT, 2014).

In some cases, at the tourism accommodation establishments, measurements are taken whilst often, the data comes from the border counts (immigration, police) which is supplemented by surveys conducted at the entry points like at the ports and airports. Therefore, all the establishments' actual coverage is reflected in the count whereas in certain cases sampling is done of areas that are peculiar and by estimation, an extension of the results is done. Furthermore, where the same people travel across a country, they may be counted several times when their stay is at differing accommodations. Eventually, following the hierarchy that is administrative comprising of all the issues related to possible missing items, wrong assessments or errors in transcriptions; the collection procedures that have been performed at certain local levels need to be aggregated (Volo, 2004). It is worthwhile to note that for both the domestic and international travels the issues are the same.

When expenditures are considered the same, it can be said with rather more problems as most of the travelers are reluctant to share details about and declare their expenses and at the same time it is difficult to distinguish as to whether certain expenses are related to tourism (Frechtling, 2006). Moreover, essentially, the consistency, overall reliability and comparability of the measurements about demand are relatively poor at the basic level leading to several challenges when comparisons and parallels are made between different countries or areas and especially when for preparing plans or making decisions forecasts are required. Hence, several characteristics over and above the usual figures on length of stay and tourist arrivals are measured in most of the countries. However, these vary from socio-demographic distributions (gender, age, education, etc.) to motivations for the trip, to economic conditions, to in-depth analyses of different geographical origins, to means used for planning and booking. Lan & Mc Kercher have reported a detailed list of the data that is available in several countries (Lam & McKercher, 2013).

At different times, by and large there is a regular collection of most of the demand metrics. An assembly of these is then done for the purpose of making predictions. Therefore, some knowledge of the developments likely to take place in the future is required foe planning, setting policies, managing, defining strategies and deciding investments at an aggregate or even individual level so that in the field of practitioners and researchers alike forecasting takes on a high importance making it the most relevant trend as far as research in tourism is concerned (Rita & Moro, 2016). Moreover, for tourism demand forecasting, a number of methods have been devised. Three major groups have been identified as of now: where for prediction of likely trends in the future historical data is use, the time series models; where the relationship between the demand and certain variables of social or economic nature is looked into, the econometric models and where the latest developments in computer science are used to apply methods like the rough sets theory, neural networks, genetic algorithms or the fuzzy-time series theory so that informative outcomes can be derived using the artificial intelligence models.

In order to forecast the situations likely to arise in the future the analysis by time series is the one adopted most to examine the tourism system's general dynamic behavior (whether it is a single operator or a destination) and for forecasting the situations that are likely to arise. This field has a plethora of proposals: exponential smoothing techniques and simple naïve models; sophisticated statistical and econometric approaches and with a varied number of variations autoregressive integrated moving average (ARIMA) which is more elaborate. At the same time many studies have gone to show most of the methods are not outperformed by a single other one and the different methods need to be combined or alternatively, certain qualitative judgments shall need to be used to revise the outcomes (Baggio & Antonioli Corigliano, 2008). Based on the machine learning principles several techniques have been made available due to the recent technological advances and developments in software. The point is that without having to program explicitly the 'what and where' aspect of looking the application should be able to learn iteratively from the data. There are basically three ways in which a machine can be taught:

- Supervised learning: Instruction is given to the machine to attempt and make its decisions by recognizing the output patterns with a set of inputs and their outputs that are given.
- Unsupervised learning: The machine finds structures and patterns in the input by itself using various algorithms with the set of inputs that are given to it.
- Reinforcement learning: by trying to maximize some function and optimizing, the machine utilizes the environment and data to learn dynamically.

5.6. FORECASTING TOURIST ARRIVALS USING TIME-VARYING PARAMETER STRUCTURAL TIME SERIES MODELS

The economic growth in many regions and countries is significantly contributed to by tourism. For tourism business practitioners and tourism policymakers the demand in tourism in the future and its trends need to be predicted accurately especially considering the rapid increase in tourism internationally during the recent decades. Furthermore, the seasonal variations play a significant role in most of the demand for tourism of a destination. In myriad ways, seasons affect tourism and the variations in these are responsible for business failures and high risk for investment; for the difficulties faced in getting access to capital, difficulties faced in maintaining a quality of service that is consistent and the utilization of facilities as well as resources in an ineffective manner. At the same time off-peak season has certain inherent benefits like "resident" recovery and gives sufficient time for the environment reclamation so that seasonality cannot always be taken as detrimental towards tourism (Butler, 1994).

However, seasonality has its own pros and cons, but it is essential that the seasonal patterns of the demand for tourism are well known in detail so that accurate predictions can be made for the future values in their regard and operations as well as planning can then be done effectively like capacity management, staffing and allocation of resources. Moreover, the effect of various factors and the seasonal patterns of demand for tourism have shown through empirical evidence to change with time. Hence, the effects of explanatory variables and seasonality both need to be modeled appropriately so that the demand for tourism in the future can be forecasted accurately.

In comparison to the deterministic models, studies on tourism demand forecasting have shown far better performance with the time-varying parameter (TVP) regression approach through, which over time relaxes the restrictions on the demand parameters' constancy and the structural time series models (STSMs) through which cycle, trend and seasonal components of a variable are specified as stochastic. Moreover, this study aims to construct an economic model that is new and by introducing the explanatory variable coefficients' TVP estimation further develops the casual STSM thereby combining the merits of the TVP and STSM model.

In comparison to the methods used earlier the forecast of the seasonal demand for tourism made by this new TVP-STSM is expected to be far more accurate. The accuracy of the proposed model to forecast is evaluated through the empirical study for the purpose of forecasting the arrival of tourists to Hong Kong from the four main markets namely: South Korea, China, USA and UK.

5.6.1. Modelling and Forecasting Seasonal Tourism Demand

Over the last two decades, a plethora of literature has been published on the analysis and forecasting of seasonal demand in tourism and our understanding with respect to the features associated with seasonal demand for tourism has been contributed to significantly through this literature. All the same, these studies have certain potential problems.

The patterns associated with the seasonality in demand for tourism was considered as constant in the studies conducted earlier. However, seasonality cannot be determined due to various changes like weather and climatic conditions, politics and technology and the popularity of tourist destinations and activities. The seasonal tourism demand studies were introduced with the STSM which though later refined by Harvey in 1989 was proposed initially by Harrison and Stevens in 1976 so that the assumptions of deterministic seasonality could be overcome. However, a time series is decomposed into its seasonal, trend, irregular components and cycle without including the explanatory variables (also known as the BSM or the basic structural model) by the basic STSM and these components are regarded as stochastic. As a result, the seasonal demand in tourism is better reflected through this model in comparison to the traditional models that were constant seasonal time series one.

All the same, effects that the economic determinants have on the interest variable are not considered by the BSM (and in general by the univariate time series models). So that the casual variable could be included in the model specification, the BSM was further developed so that this limitation could be overcome (known as the CSM or the casual structural model). Moreover, CSM and BSM have both been applied in the context of tourism demand forecasting and relative to the other alternatives in the time series their superior performance in forecasting has been demonstrated in many studies (for instance one can see Turner & Witt, 2001; Witt & Kulendran, 2001; King & Kulendran, 1997; Moral & Gonzalez 1995, 1996). However, CSM was seen to produce forecasts that were less accurate relative to the BSM in studies of both Kulendran and Witt (2003) and Turner and Witt (2001).

Nevertheless, CSM relates to the explanatory variables' treatment, and this could be the possible reason for the forecasts made by it not being as accurate as the ones made by BSM. The explanatory variables' parameters are taken to be constant over a period even though seasonality, cycle and trend in the CSM are all taken to be stochastic. Therefore, the implication drawn from this is that there is no change in the economic structure that generates the data. A double-log form specified demand model, over a period, the elasticity in the demand for tourism is constant as suggested by the constant parameters and this is quite unrealistic as well as restrictive.

However, people may in fact react differently as per the changing economic environment to a stimulus that is given at varying points of time. A stochastic process is likely to be followed by the changes in the coefficients due to the ambiguous or transitory nature of the modifications to the environment (Lucas, 1976). This is however not considered by the CSM which could be the likely cause for its forecasting performance being unsatisfactory. Moreover, the late 1990s saw the introduction of the time-varying parameter (TVP), with modeling approach within the context of tourism so that the above limitations of the demand models with the traditional dimension of fixed-parameters could be overcome.

The restriction on the explanatory variables' constancy of the coefficients was relaxed by this approach allowing the stochastic parameters to better reflect the elasticity in demand and its evolution over a period. Compared to the other econometric models more so in the short term, more accurate forecasts were generated through the models using the TVP approach as can be seen through various TVP technique based empirical studies conducted previously. For instance, the performance of the TVP model vis-à-vis the other economic models with fixed parameters and time series models, for forecasting international tourism was examined by Song and Witt (2000); Song, Witt, and Jensen (2003) and Witt, Song and Louvieris (2003).

In terms of the error magnitude whereby the root means square percentage error (RMSPE) and the mean absolute percentage error (MAPE) measure the forecasting accuracy was assessed in the first two studies. In the forecasting comparison involving one-year-ahead, the TVP model was shown to outperform all the competitors in both the studies wherein these other models included the ECMs, vector autoregressive (VAR) model, the autoregressive distributed lag model (ADLM) and the naïve no-change model. In terms of the directional change and error magnitude, the TVP models' forecasting performance was investigated by Witt et al. (2003). In a competition held between seven candidates, in both the assessments the TVP model was shown to rank the second best in the comparison for a year ahead forecasting.

For a short run, forecasting and more specifically for one year ahead, the TVP technique has been ranked second or first in all three empirical studies implying that the suitability of the TVP model for the purposes of tourism's short-run planning is high. Therefore, the data for the annual tourism demand is however used by all the applications above of the TVP model and these studies did not examine the influence that seasonality has on the demand for tourism. In order to forecast the demand for seasonal tourism, seasonal dummies can be easily incorporated in the above models (Shen, Li & Song, 2008 can be seen). However, deterministic seasonality shall be implied by this.

For the purposes of forecasting the seasonal demand for tourism, for the first-time attempt was made by the current study to get the structural time series models incorporated with the parameters that are time-varying. Forecasting accuracy is likely to improve through the seasonal demand's comprehensive analysis provided by this modeling approach which is new. For the demand for tourism in Hong Kong from four main markets of source to forecast the quarterly demand, application of the TVP-STSM is made. For the key players in tourism quite useful information shall be provided by this study's empirical results whilst it shall also help the public agencies to formulate policies for tourism and evaluate their effectiveness.

5.7. AN INTEGRATIVE APPROACH TO TOURISM FORECASTING

For the state of South Australia in Australia, forecasts of tourism for the year 2005 were made through the application reported herein. Mass destination is not the forte of South Australian tourism. In order to attract tourists Murray River associated "big river" theme and the wine regions (like Clare Valley, McLaren Vale and Barossa Valley regions) were heavily relied upon by the state as reported by Faulkner, Oppermann and Fredline in 1999. Only seven percent of the visitor nights by domestic Australians and four percent of the visitor nights by international tourists were accounted for by the state. All the same, tourism is regarded as a growth industry by the state authority for tourism (South Australian Tourism Commission or SATC) and for the state's destinations, there is a keenness to capitalize the economic potential with regard to tourism's growth leading to the forecasting process' value. Moreover, forecasting helps to minimize the disparity between supply and demand related losses for the operators as future is attempted to be anticipated through it wherein its value for many business endeavors lies. Tourism requires that the future demand for accommodation, transport, retail, skilled labor and various facilities for entertainment are essentially anticipated well in advance. On the other hand, several years beforehand, the future demand should be anticipated as the development of infrastructure requires the involvement of a large amount of time and needs to be ensured that the market can be well serviced. Amenities shall be lacking, congestion shall result, services shall be poor and due to missed market opportunities there shall be the inevitable los if market share if the future demand for the tourism destination is underestimated as alternative destinations shall be sought by the tourists. On the other hand, there shall be an excessive supply of infrastructure, resources shall be used inefficiently and the return on investments shall be low if the future demands are overestimated. Moreover, the confidence of the investor and perception of the visitors shall be damaged considerably if market forces decide upon the provisioning of infrastructure for tourism due to the cycle of excess supply following the excess demand that shall result from it.

Figure 5.5. Integrative approach to tourism forecasting.

For all destinations and all forecasts, no approach is likely to provide forecasts that are accurate for tourism forecasting. A number of statistical approaches towards forecasting tourism were reviewed by Witt and Witt (1992) which included the time-series modeling and detailed econometric modeling. Moreover, qualitative forecasting methods like the expert opinion technique or the Delphi were specifically excluded from the course of their study. All the same, this qualitative forecasting method has been included in other research studies (Uysal and Crompton, 1985; Calantone, di Benedetto, and Bojanic, 1987).

The overview of the integrative approach are as follows. A combination of qualitative and quantitative methods for forecasting as advocated by Frechtling (1996) have been used to put into perspective the overall approach that has been adopted by this study. This study particularly used a Delphi panel, single exponential smoothing forecasts and naïve forecast in combination.

5.7.1. The South Australian Forecasting Approach

The Table below shows the segments as per origin into which the tourism into South Australia was put. To the past data, six varying time-series methods for forecasting were applied for each international segment so that the number of likely visitors up to the year 2005 was generated through six different forecasts. Therefore, the exercise was conducted in the year 1995 and all the years after this were considered as forecasts. The annual data from 1984–1985 with respect to the number of visitors was used for the international tourists visiting the state. In the table mentioned below each of the segments of origin were applied with one "top-down" approach and three time-series approaches which were quite simple. The time-series approaches used were as follows:

Nevertheless, Holt's exponential smoothing: The average changes taking place in the long-run trend were incorporated in this method of time-series (Pindyck and Rubinfeld, 1991). The data's relatively recent trends were emphasized more with this model, as a result of which the latest trends were reflected the most in the forecasts that thus resulted. Moreover, in the future too based on the assumption that there shall be a continuation of the average annual rate of change from the previous eleven years, a naïve method would continue. Hence, by utilizing regression analysis, the annual data shall be fitted with a simple linear trend. It is assumed in the forecasting model here that each time there shall be an increase in the series by an absolute amount that is constant. The average change that takes place in the actual data set from the first to the last point is the basis for such an increase.

The Australian Tourism Forecasting Council (TFC) generated a forecast of the overall number of tourists that shall arrive in Australia and the top-down approach used this forecast as well as applied the percentage to which each originating market was contributed to by the Australian tourism. Moreover, industry practitioners modify the economic-demand model derived forecasts through a process of the TFC and the forecasts of the number of international tourists traveling to Australia is published by the TFC.

The forecasting methods in the study relied on the median shares of visitation during the relatively recent times-based share of South Australian market in the national forecast even though historical data with respect to eleven years from 1985–1995 was available for South Australia. Moreover, this time frame was considered to provide a far sound base for the purposes of the forecasts as it reflected the trends that were more recent. The highest, median and lowest share of the data for the last seven years formed the basis for the three different percentages for the share of South Australia towards the national total.

The Delphi instrument included tables like this for every segment of international origin. Moreover, the forecasts have been shown through the numbers in italics and the actual numbers of tourism for the segment from 1984 to 1995 is shown in the table's top half. It may be noted that no data collection was there for 1987. For the required stat-of-origin basis there were no TFC forecasts available for the segment of domestic market, as a result of which the top-down approach could not possibly be used. Hence, for the domestic market too, the three time-series methods that were used for the international market were used. However, it was felt that the number of visitors would be better related to by the people from the tourism industry so not using the visitor nights, visitor numbers were used in all the forecasts. In order to devise the expenditure parameters more readily the forecasts would need to be converted to visitor nights as extension of this work so that estimates of the economic impacts on the growth of tourism could be produced (Mules 1998).

5.8. DIFFERENT METHODS OF FORECASTING

Forecasting can basically be of different variations like:

- Seasonal or short term;
- Annual or mid-term;
- More than 2–5 years or long term.

To suit the set-up of the industry the methods used for forecasting tourism must be applied even though these methods are like the ones used for the other products' forecasting. These can be as enumerated below:

- Experience based judgment,
- Time-series,
- Surveys,
- Market tests,
- Correlation.

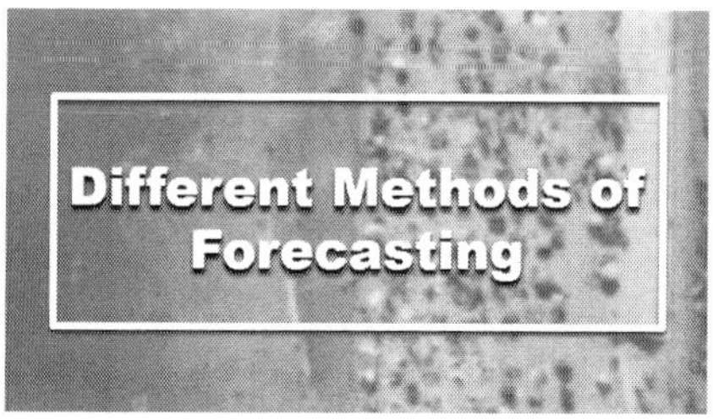

Figure 5.6. Different methods of forecasting.

5.8.1. Survey

Small private firms or governmental organizations can conduct survey. The preferences of the tourists as well as their impression with respect to the services in tourism can be found with the help of these surveys. At the international airports' departure lounges a survey is carried out by a team of the DOT wherein the visit of the tourists is completed, and they can answer the survey-based questions with time in hand. Moreover, in order to understand the opinion of the customer's, feedback is taken continuously by hotels and airlines which helps to design products. To predict product's quality and quantity several surveys can be conducted for sales force, experts and customers.

5.8.2. Time-Series

The seasonal fluctuations in terms of the highs and lows can be understood better by the utilization of historical data and then on time basis, sorting it and the tourism industry can benefit from these kinds of details as it is a very seasonal industry. This is essential to be prepared for the high number of tourists in the peak season and in the low season, for the promotion of sales figures.

8.5.3. Correlation

Variable factors like the per-capita incomes can be linked with by the tourist flows with the help of this method. Furthermore, this is also essential for the working out of the marketing plan as well s the mix of product, place of distribution, price and promotion.

5.8.4. Market Tests

There is the conduct of market tests to understand the competitors' products and the market demand. Sure, and confirmed clues of pricing, product design etc. can be given through this. The judgment of senior partners and senior executives based on their experience is a commonly used method. Detailed technical analysis is not required in this method also, it is very quick, and these are the main advantages of this method. However, at times, this method may not be quite reliable. Hence, for the same period, different forecasts may be produced through different methods of forecasting. Based on the specific situation for which forecasting is to be done an appropriate method for forecasting pragmatism is reassured.

5.9. APPLICATIONS IN TOURISM

Tourism has certain peculiar characteristics hence forecasting and the application thereof becoming a necessity for it even though the same is quite important for planning of any marketing. Stocking of whichever product is not possible is essentially perishable. Accuracy in marketing planning can be brought about only through forecasting that is accurate. Moreover, there should be systematic planning of places of various factors like promotion, product designing and pricing methods for the organizations that operate tours. Before deciding, the number of competitors and the sector's popularity need to be considered by a tour operator whilst the sector of tour is being selected. Similarly, for such a business it is essential to have forecasts about the tour package's likely buyers.

Quality and pricing are other aspects that need to be taken into consideration. The market trends pertaining to the Indian market are seeing rapid changes as a result of the government's liberalization policy. International norms are being followed for the dual aspects of quantity and quality hence is essential that for them to be fully effective that they too are brought up to the global standards. Moreover, early bookers are often offered special rebates by tour operators e. g. using a lucky draw to lure customers. Understanding the trends in pricing, product design and promotion has become essential for survival in the strong competition. Hence, the organizations may land up with huge losses if the rebate and discount norms are not related to the sales figures.

Strategies activities like diversification, expansion and merger, etc., too require forecasting. The organization's performance falters badly when the forecasts are not accurate whereas the success rate goes high when these very forecasts are made accurately. Novel and ingenuous methods are worked out to ensure that even in the low season the requisite minimum sales figures are met in tourism. Special rates and high discounts are offered by hotels and resorts for seminars and conferences that are booked during the non-season periods. Relatively lesser fluctuations are seen in business tourism in comparison to the other areas as a result they can be used to cover the gap. Furthermore, the planning of various organizations is different in anticipation of a high tourist influx during a season. Hence, proper demand patterns need to be projected well in advance as resorts need to make temporary arrangements at times in terms of tents for accommodation and additional services need to be run by the transport operators.

Projection of demand is the most important facet of application along with organization's capacity for production of service and thereafter planning towards getting the possible share of the targeted market. In order to get predictions that are more accurate historical data should be used as a premise for evaluating the forecasts. To further understand the accuracy projected figures should be checked vis-à-vis the feedback.

5.10. CONCLUSION

Tourism industry is growing rapidly throughout the years, and it is contributing to the large part in the world economy. Tourism industry is playing a vital role all over the globe. Therefore, an estimate and forecast of the tourism and transportation needs to be done which is discussed briefly in this chapter along with various forecasting methods and the different types of the time series models. Moreover, it gives an overview of the four-step travel model process and an insight into the several tools for measuring tourism like the methods, indicators and needs. Subsequently, it discusses about the current issues and methods in the tourism industry, forecasting tourist arrivals using time varying parameter structural time series models and talks about an integrative approach to tourism forecasting. Hence, its numerous methods such as survey, time series and correlation. The last topic of this chapter, throws elucidate the applications in tourism, and how it is prerequisite in the tourism industry.

REFERENCES

1. Baggio, R. (2018). *Measuring Tourism: Methods, Indicators, and Needs*. [ebook] Available at: https://www.iby.it/turismo/papers/baggio_MeasTourism.pdf (accessed 8 August 2021).
2. Chandra R. Bhat, Cambridge Systematics, Inc, Vanasse Hangen Brustlin, Inc., Gallop Corporatio, Shapiro Transportation Consulting, LLC and Martin/Alexiou/Bryson, PLLC, (2012). *Travel Demand Forecasting: Parameters and Techniques*. [ebook] Washington, D.C: Transportation Research Board. Available at: https://transportation.ky.gov/Planning/Documents/Travel%20Demand%20Forecasting.pdf (accessed 8 August 2021).
3. Ghalehkhondabi, I., Ardjmand, E., Young, W., & Weckman, G., (2019). A review of demand forecasting models and methodological developments within tourism and passenger transportation industry. *Journal of Tourism Futures,* [online] *5*(1), pp. 75–93. Available at: https://www.emerald.com/insight/content/doi/10.1108/JTF-10–2018–0061/full/html#sec003 (accessed 8 August 2021).
4. Haiyan Song, Gang Li, Stephen F. Witt, & George Athanasopoulos, n.d. Forecasting tourist arrivals using time-varying parameter structural time series models. [ebook] Monash University, Australia. Available at: https://core.ac.uk/download/pdf/17029229.pdf (accessed 8 August 2021).
5. Petropoulos, C., Nikolopoulos, K., Patelis, A., & Assimakopoulos, V., (2005). A technical analysis approach to tourism demand forecasting. *Applied Economics Letters*, [online] *12*(6), pp. 327–333. Available at: https://www.tandfonline.com/doi/abs/10.1080/1350485050006574 5?journalCode=rael20& (accessed 8 August 2021).
6. Tideswell, C., Mules, T., & Faulkner, B., (2001). An Integrative Approach to Tourism Forecasting: A Glance in the Rearview Mirror. *Journal of Travel Research,* [online] *40*(2), 162–171. Available at: https://www.researchgate.net/publication/37357819_An_Integrative_Approach_to_Tourism_Forecasting_A_Glance_in_the_Rearview_Mirror (accessed 8 August 2021).
7. Unit 6 Forecasting For Tourism and Its Products. n.d. [ebook] Available at: https://hmhub.me/wp-content/uploads/2017/11/Unit-6-Forecasting-for-Tourism-and-its-Products.pdf (accessed 8 August 2021).

Chapter 6

Challenges and Opportunities in the World of Tourism

CONTENTS

In the chapter, challenges and opportunities in the world of tourism, the importance of sustainable tourism as a part of development strategy is discussed. It also explains the role of sustainable tourism in poverty alleviation in third world nations. In addition, it also shed some light on the importance of travel & tourism in creating livelihoods and opportunities for everyone in society. Moreover, it explained the meaning of big data and how it can be used to get the maximum from tourism sector. Hence, it discusses several problems in capacity building of sustainable tourism development. It further highlighted the growth paradox: can tourism ever be sustainable. At the end, it explains several problems in providing tourism infrastructure.

6.1. INTRODUCTION

The travel and tourism industry has appeared as one of the leading and fastest growing economic sectors across the world. According to a report by UNWTO (2013), it was noted that tourism's total contribution to worldwide GDP is projected at 9%. Exports of tourism in 2012 stood to USD 1.3 trillion accounting for 6% of the total world's exports. New tourist destinations, particularly those in the emerging markets have started adding prominence with traditional markets reaching maturity. Asia Pacific logged the highest growth in the sum of international tourist arrivals in 2012 at 7% accompanied by Africa at 6%. Progressively, travel and tourism are evolving as an imperative category of services exports worldwide.

The role of travel and tourism sector become even more important in countries that are developing as it provides them several socio-economic benefits. Generation of revenue, and foreign exchange, provision of employment, expansion or development of other industries such as construction, agriculture, handicrafts etc. are some of the important economic benefits provided by the tourism sector in the developing countries. In addition, ensuring investments in infrastructural facilities such as accommodation, transportation, and other tourism related services results in an overall development of infrastructure in the economy. Moreover, it is well known that Travel and Tourism is a crucial for human beings. It is hence a process and situation, where person from one place is visiting the other place and country for a shorter period. This concept is known as tourism. However, nowadays, the tourism industry carries a greater significance. No one can deny the number of benefits that are provided by the tourism sector. Therefore, a foreign tourist visited to a home country provide greater foreign exchange earning to the country.

Some of the important feature of incorporating tourism policy in a country are discussed as follows:

- Tourism is one of the most important tools for employment generation, rural transformation and economic development.
- To take benefits of global trade transaction through travel and tourism.
- This policy is dependent on seven key indicators of tourism development. These indicators are welcome; facilitation; information; safety; infrastructural development; Co-operation; cleanliness.
- To use natural resources, human resource, and technical resources for sustainable development.
- To use labor intensive technique in tourism sector for improving standard of living by employment generation.
- To focus on rural areas for low-cost programs related to tourism centers.
- To create forward and backward associations in the tourism sector for overall development.
- To surge the foreign earnings through export of tourism services.
- To promote understanding, harmony and to contribute regional stability and national unity.
- To develop shopping centers for generating revenue and other rural tourism products.

6.2. SUSTAINABLE TOURISM AS A PART OF DEVELOPMENT STRATEGY

Figure 6.1. Conservation in tourism.

Source: Image by Routes online.

It is generally seen that third World countries are particularly interested in international tourism, and majority of people have opinion that it brings countries a large selection of economic benefits including small business development, employment opportunities, and increased in foreign exchange reserves. Most of them believe that more money can be gathered through developing luxury goods and services although this rises a countries reliance on foreign investments, imported products, and expatriate skills. This classic 'trickle down' financial strategy hardly makes its way down to advantage people at a grassroots level. Moreover, it is believed that the economic benefits of large-scale tourism cannot be questioned but that the budget traveler or backpacker sector is usually neglected as a potential growth sector by Third World governments. Hence, this sector brings noteworthy non-economic benefits that could assists in empowering and educating the communities embraced in this sector.

Nevertheless, "Aiming 'low' builds upon the skills of the local population, promotes self-reliance, and boost the self-confidence of community members who are dealing with outsiders, all signs of empowerment" and all of which aid in the overall development of a nation. Therefore, by now that we have included the concept of sustainable tourism, one may well infer what exactly is the point of attaining the principles behind the aforesaid. For example, in Maori, there is a famous proverb – Toitū he whenua, whatungarongaro he tangata. In other words, it can be said that land is permanent while man is not. Thus, by now, people are living in an era where there is an increasing concern about environmental pollution and climate decimation. At the same time, countries, especially the developing ones are trying to forge a distinct identity of their own in the international arena. Unhappily, to mark such an identity, there is huge to make its presence felt and, to make such a presence to be felt globally, there needs to be progress in such said nation of the Third World cadre. Therefore, in this section, the focus will be on how sustainable tourism is essentially the best of the bets such nations have to harvest more and more international awareness. Further, it is equally important to discuss about how an investment in the tourism sector is far smarter an idea in comparison to encouraging growth in the technology-based economy sector.

6.2.1. Sustainable Tourism Against Technology Based Economic Sector

According to several reports published by the World Tourism Organization, in the year of 2000 alone there was an enormous expenditure of over USD

478 billion by over 700 million people traveling to a foreign nation for sightseeing and tourism purpose.

These figures do not include the visits made to a foreign land for business meets and personal meet. In addition, international tourism receipts added with passenger transport add up to over USD 575 billion. Therefore, according to figures and facts of circa 2000 AD, tourism tops over chemical exports, automotive products, and petroleum exports and food trade, making it the number one earner. The above-mentioned facts and figures follow only to the tourism export business. In this, we are not even including the income that would be made due to the vast employment generation. Hence, by now, that was all about how tourism as a sector of economic development is a forte. When rivalled to technology-based-economic sectors, some other valuable factors should also be taken into consideration.

On the other hand, sustainable tourism is basically concerned with giving people a closer look at the nature that they have been ignored majority of the times. Partly, due to the scarcity of time people have in their life to associate with it and partly due to the spots for such tourism are not in copiousness, anymore. Moreover, environmental pollution has, to a very greater extent, guzzled down a lot of scenic spots, leaving such spots in a dreadful condition. If one vigilantly looks at the facts and figures, it is easily observable that most of the tourists who travel to such distinctive lands are from First World nations.

However, one of the key reasons behind this is that the automated lifestyle that people in such First World nations usually lead is monotonous and mechanized, usually limits and the corporate work culture typically eats up most of their time. Therefore, markedly, the desire for a break from such a lifestyle trigger them to travel to places which have that aesthetic and scenic and picturesque value to them. Basically put, the developing countries, which proposes a lot of serenity and beauty. This, in turn, inspires such economically backward developing countries to promote themselves as a tourist destination. This system of tourists from foreign lands visiting and such developing countries becoming hosts works on the basic beliefs of demand-and-supply. And, in turn, the tourists also meet the ethnic and local communities of the Third World nations, above-mentioned. This, also, open the pathways for a higher understanding between cultures and people. Such understanding, eventually, helps in the process of harmonizing nations and the communal will of people – which, for over a period, helps foster a sense of brotherhood and fraternity.

6.3. THE SPECIAL IMPORTANCE OF SUSTAINABLE TOURISM IN POVERTY ALLEVIATION IN THIRD WORLD NATIONS

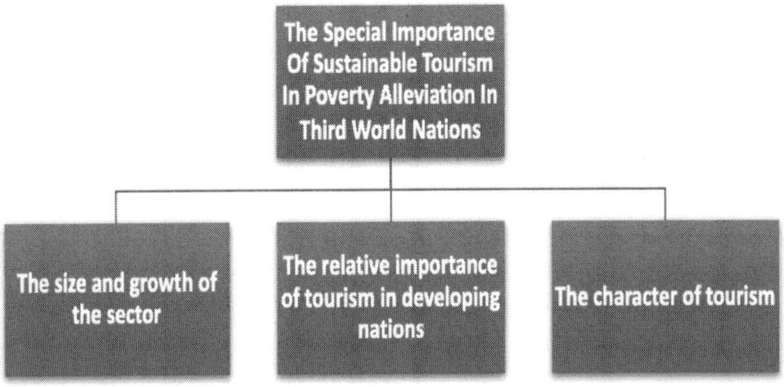

Figure 6.2. The special importance of sustainable tourism in poverty alleviation in third world nations.

Taking the idea of overall development of people of developing countries one step ahead, here is how sustainable tourism can play a very crucial role when it comes to poverty alleviation of people and improving the standards of living of people.

6.3.1. The Size and Growth of the Sector

It is generally seen that in several countries, tourism acts as an engine for development through earning of foreign exchange and generating direct as well as indirect employment. Tourism contributes to almost 5% of the world's GDP. It accounts for 6% of the world's exports in services being the fourth largest export sector after chemicals, fuels, and automotive products.

Tourism is also accountable for 235 million jobs or one in every 12 jobs worldwide. In 2011, it was estimated that international arrivals grew by over 4% reaching 982 million, up from 939 million in the year 2010, in a year portrayed by a stalled global economic recovery, major political transition in the North Africa and Middle East and natural disasters in Japan.

6.3.2. The Relative Importance of Tourism in Developing Nations

It is generally seen that tourism in many developing and least developed countries is the most sustainable and viable economic development option, and in few countries, the chief source of foreign exchange earnings. Part of this income trickles down to distinctive groups of the society and, if tourism is succeeded with a main focus on alleviation of poverty, it can directly offer assistance to poorer groups through employment of local people in tourism enterprises, goods and services engaged in tourism sector, or the running of small and community-based enterprises, etc., having optimistic impacts on lessening poverty levels.

Tourism in the past few years has been illustrated by two chief trends; firstly, the consolidation of conventional tourism destinations, such as those in North America and Western Europe; and secondly, a noticeable geographical expansion.

There has been a considerable variation of destinations, and several developing countries have seen their tourist arrivals increase considerably. Arrivals to developing countries accounted for 46% of the total international arrivals in the year 2011. It has been observed that tourism has become a key player in the economy of developing countries.

Some important facts are discussed as follows:

1. In 2011, international tourism arrivals to emerging market and developing countries accounted for almost 459 million.
2. Tourism accounted for major proportion of export earnings in 20 out of world's 48 least developed countries.
3. In some developing countries, especially small island states, tourism can account for over 25% of GDP

6.3.3. The Character of Tourism

There are several characteristics of tourism as an activity that make it specifically pertinent to low-income countries, and to poor communities within them. These comprise:

- Its response to particular assets – It is generally seen that tourism places significant value on some common features of developing countries, such as rich cultural heritage, warm climate, abundant biodiversity and inspiring landscapes.

- These strengths can be principally evident in rural areas, which may have a relative benefit for tourism while being at a shortcoming in most other economic sectors.
- Its accessibility to the poor – tourism is a comparatively labor-intensive sector and is conventionally made up of small and micro enterprises. Majority of activities in tourism are chiefly suited to women, young people as well as disadvantaged groups for instance ethnic minority populations.
- Many tourism jobs are possibly quite reachable to the poor as they need comparatively few skills and little investment. Some may also be part time and used to adjunct income from other activities.
- Its connectivity – as so many distinctive activities and inputs make up the tourism product, which has an extensive and varied supply chain, spending by tourists can offer advantage to array of sectors such as handicrafts, agriculture, transport as well as other services. Additional rounds of expenditure by those people whose income source are from tourism disseminate the economic benefit further (the multiplier effect).
- Its linking of consumers to producers – tourism, infrequently, is an activity that gets the consumers to the producers. The communication between tourists and poor communities can offer an amount of intangible and real-world benefits. These can vary from increased awareness of environmental, cultural, and economic issues and values, on both sides, to mutual benefits from value-added local investment in infrastructure sector.

6.4. TRAVEL AND TOURISM CREATES LIVELIHOODS AND OPPORTUNITIES FOR EVERYONE IN SOCIETY

It is worth noticing that tourism is one of the world's biggest employers. Almost 300 million people earn a living from the sector directly as well as indirectly. That comprises almost 10% of all jobs on the planet.

In fact, it is generally seen that at present, there is no other sector that offers employment to so broad a section of society. Therefore, what is more important to consider is the types of people working in these jobs.

Figure 6.3. Travel & tourism creates livelihoods and opportunities for everyone in society.

Source: Image by unsplash.com.

6.4.1. Travel and Tourism Normally Employ Higher Percentages of Women

Evalyn's story is one of many. For instance, women work in 65% of the tourism and travel jobs in Australia and Germany in comparison to 45% for other sectors in these countries. There is still existence of huge pay gap between men and women, but it's typically far smaller too.

It is generally seen that travel and tourism tends to employ young people. It was noticed that almost half of all employees in the hotel, hospitality, and catering sector are under the age of 25 years. This is especially imperative for economies in the less urbanized world where the young people percentage is far higher.

6.4.2. Travel and Tourism Offer Jobs to People with Little Formal Training

By its very nature, the tourism industry is often labor intensive, greater proportion of people to prepare food, providing room service, and maintain infrastructure of hotel. These are jobs that are comparatively safe from the advance of artificial intelligence and technology.

6.4.3. Travel and Tourism Offer Employment in Far-Flung Places

As the growing tourism trends for uncovering unmapped destinations and having meaningful experiences continue, the tourist impact spends spreads to more far-flung parts of the planet. Take tiny landlocked Bhutan that is following a careful sustainable tourism development plan. Tourism arrivals have risen from mere 20,000 a year to almost 160,000 in the last seven years, fetching over 30,000 potential jobs along the way.

6.4.4. Travel and Tourism Substitute Illegal Jobs with Legitimate Sources of Income

It is not all about just about providing opportunities to people who might else struggle to find work. Travel and tourism can also offer sufficient alternative income to people who earlier dependent on dangerous or illegal work for their livelihoods.

Figure 6.4. Travel and tourism substitute illegal jobs with legitimate sources of income.

Source: Image by unsplash.com.

6.5. WHAT IS BIG DATA AND HOW IS IT USED?

Big Data that can be defined as the data in large volume, plays a very crucial role in analyzing the distinctive trends of travelers by embracing the information attained via several consumer centers and employing it to establish a definite marketing approach and strategy that can be applied to the target audience. Moreover, these technologies that comprise of MongoDB, Hadoop, cloud-based analytics, and Rainstar, offers sufficient space for data storage. They therefore present the information amassed from a large range of sources in a structured and organized manner that supports the businesses involved in the tourism and travel industry to engage in active decisions making in relation with inconsistent and varying consumers needs and wants.

One of the popular and useful method is Website data scraping. In this process, the information gathered from several websites is transformed into a raw data format such as '.csv' or a text file which is then fed into models for data analysis purpose. With this, the data scraped from public websites can be taken into use in order to collect useful insights and information for creating a product portfolio that can be released into the market later. However, another realistic and practical achievable method of the same would be Social Media Analytics. Some views and opinions of the general customers with respect to a specific brand of the company are analyzed and gauged.

6.5.1. Utilization of Big Data

Big Data has been effectively and efficiently integrated in a variety of areas in tourism. These comprise:

- **Travel bots:** It is a recently introduced feature, travel bots are chatbots that offer either automatic customer services on travel companies websites or operate through messaging platforms such as Facebook Messenger or WhatsApp to interact and communicate with travelers and assist them in their bookings.
- Powered by **Artificial Intelligence**, reinforced by multiple languages, and accessible 24*7 the Travel bots have the potential to answer queries, working in an efficient and effective manner and saving both money and time of the user, managing the trip and even of recommending suggestions.

- One of the perfect examples of a travel bot is KAYAK which is available on iMessage, Amazon Echo, and Messenger. KAYAK united with Alexa Skills helps users in tracking their flights, hotels bookings and looking for several holiday opportunities.
- **Personalized Marketing:** As it is well known that the target audience being so mixed and diverse, it become even more important to make proper use of Big Data.
- **Assists in Understanding the Demographics:** It is generally seen that Big Data assists in understanding the demographics of the target audience along with the behavioral, geographical, and psychographic factors intricate in order to recognize marketing opportunities as well as frame an effective marketing strategy. This allows for more personalized and targeted promotional content to be dispersed.
- **Boosting Customer Experience:** The Big Data when applied in a proper way can help in improving the overall experience of the customers. Data gathered from customers with respect to their sentiments, behavior, pattern with respect to that brand range between conversations on social media reviews uploaded online as well as the data attained of services used. This info therefore gives lucidity on which services can be used fully in order to extract the maximum benefits, which get used rarely, and which ones are likely to be sought. Thus, this data opens the pathways for companies to comprehend the areas in which investment is required as well as the potential services that they can introduced.
- **Optimization of Route:** Planning of a trip while considering several factors such as the destinations, the traveler schedule, his/her working hours as well seeing the distance can get testing. This is where Route Optimization comes into play. Its mono objective becomes lessening the cost and distance of traveling as well as handling the time of the journey efficiently in order to fully please the customer.
- **Predictive analytics:** The price-fixing can be finished by considering a number of factors such as the weather conditions, availability of places, season, rooms and seats.
- **Weather conditions** can be assessed effectively by taking into consideration big data analytics. It can make use of past patterns and offer future weather forecasting.

Figure 6.5. Airlines: one of the various tourism divisions which incorporate big data include.

Source: Image by unsplash.com.

Self-learning algorithms become a tool for accruing and amassing data from the past, taking the peripheral factors into consideration in a way to assess the price fluctuations in the future. Moreover some duties which these analytics is aiding with are certifying that there is consistency in the details being offered on official websites of organization and to the third-party providers for booking purposes, in cutting down on discounts for the days where exceptional promotion is not desirable, and in swelling the rates during the weekend.

6.5.2. Various Tourism Divisions Which Incorporate Big Data

- Airline: It was noted that Big Data enables the airline operators to grow an understanding of the behavioral patterns of passengers, to pinpoint the mode of travel and time that they prefer the most, to develop an understanding about the overall functioning of the industry, broaden their network connectivity capacity and help in managing the revenue.
- The Big Data plays very crucial role in determining the pricing approach to be integrated and in classifying the several emerging trends in the industry to choose the suitable reaction.

- Hotel chains: Big Data has proved to play an eminent role in the upgrading of hotel services. It is used by hotel chains in a way to form customized packages, to offer several travel-specific discounts and for offering add-on services.
- One of the numerous examples that highlight the application of Big Data analytics in hotels is the international hotel chain of Starwood Hotels and Resorts that make use of Big Data analytics by utilizing the information on macroeconomic factors and local elements and factors to establish a strong pricing strategy and policies for raising the revenues.
- Tourism Boards: Big Data is also being incorporated by tourism boards to comprehend the overall performance of the tourism industry and to get into notice the areas where additional investment is needed to create more opportunity.

6.6. PROBLEMS IN CAPACITY BUILDING OF SUSTAINABLE TOURISM DEVELOPMENT

Figure 6.6. Sustainable tourism.

Source: Image by Wander Wisdom.

Over a period of time with the increase in the number of tourists, it become even more imperative to think towards development of sustainable tourism. It is necessary for the preservation as well as protection of geographical, historical, natural, and cultural heritage sites. Though, it is not as easy-going as it appears to be, there are number of problems that the tourism sites organizer faces in ensuring the sustainability of these sites.

Each country has its own sets of problems in sustainable tourism sector as each varies in its own level of development, touristic characteristics and national development priorities, strategies and policies. However, some of the problems that are common in many countries are described as follows:

- **Lack of inherent tourism potential (natural, historical, cultural, etc.)**. In some countries, it is often seen that development of tourism is impeded by their lack of inherent natural, historical or cultural tourism resources. However, as mentioned earlier, though it is a crucial factor, tourism heritage assets alone cannot make a successful tourism industry.
- **Lack of knowledge and awareness**. One of the key reasons that act as a roadblock in the sustainability of tourism sector is lack of knowledge and awareness about tourism. In some of the countries, there are still lack of information and comprehension about economic prominence of tourism as an industry and its crucial role in offering employment opportunities to local people.

Furthermore, there is no awareness about how it helps in accumulation of foreign exchange, which is important for country's growth and development.

- **Lack of technical know-how and weak promotional activity**. Even when there is awareness among people, in most instances, there is generally a lack of tourism knowledge and professionals. Tourism, as a modern industry, is recent in the world. It is also a new and unfamiliar activity in tom many people and tourism organization.

Another related problem is the lack of tourism promotion and mass media exposure due, in many cases, to the limited technological services available.

- **Lack of tourism-related infrastructures**. It is often seen that many countries lack the necessary infrastructure, which is critical in the development of a successful tourism sector. Primary amongst these is lodging and hotels services, and transportation, safety and security, communication services, as well as tourism information services. This makes it almost challenging to offer the standards of facilities and services which tourists require.

Furthermore, now a day, expectation of tourists surged manifold with respect to quality of services. Thus, to provide them same, it is essential to have adequate infrastructure.

- **Lack of tourism investments**. Investment in every business is essential, whether it is tourism or any other kind of business. To ensure the sustainable tourism development, it is important to have appropriate level of investment.

It is widely believed that investment is not a problem in developed countries, while it is greatest nightmare for developing countries. Investment projects in developing countries is generally viewed as high risk taking. It is a challenging task for many poor and least developed countries to gain access to reasonable financing for their touristic projects even when they manage to address the problems of project identification and planning.

- **Lack of consistent tourism strategies and policies**. There are several countries that still face troubles in getting integrated tourism policy-making due, in general, to policy conflicts between private tourism agencies and government department. In majority of the cases, this is coupled with the lack of effective administration, institutional and regulatory frameworks of touristic activity.
- **Lack of tourism diversification**. For numerous years, international tourism has been largely known for its interest in coastal or watercourse areas, and in summer for relaxation, leisure activity, and entertainment. Though, modern touristic activity has shown extensive interest in diversity of activities and programs.

This makes it problematic for many countries, including those with a more developed tourism sector, to keep pace with the speedily evolving and complex requirements of tourists. In a highly competitive international tourism market, and development of widely new touristic destinations, it is not tranquil to ensure the advancement in the conservation of tourism sites that foster modern tourism development.

- **Lack of tourism safety**. The safety of tourists is one of the imperative factors for any successful tourism industry and thus, should always be one of the basic objectives of tourism planning and provisions. It is usually perceived that safety related tourism issues and challenges exert a negative impact on the reputation of host countries.

In such a manner, negative perceptions play a very crucial role in number of tourists visiting a specific tourist country. Even such factors as rumors may cause great damage to complete tourist seasons. So, tourism safety should always be the topmost priority among host countries to create positive image among tourists. It is usually believed that good tourism safety results in silent promoting and marketing of country, thus helping in enhancing tourism base.

6.6.1. Barriers of Capacity Building in Types of Tourism Activities

Harrill and Potts (2003, p. 233) believed that "tourism is an invisible industry, encompassing transportation, loading, and entertainment. Unfortunately, tourism is also invisible to many planners, so tourism development is often left to private developers and leisure service providers."

Within the past several years, there has been substantial increase in the interest in tourism (Galston & Baehler, 1995). Ivanovic (2009) states that there are numerous types of tourism activities that play an imperative role in increasing tourism expenditure and the length of tourists' stay in communities (Ivanovic, 2009).

Despite the several possible ways in community capacity building in tourism development, there are still various barriers and challenges that are required to be tackled in planning of tourism development.

It is seen that community capacity building in tourism development always faced barriers (Moscardo, 2008). Moreover, these barriers are often overlooked by scholars and city planners in debate of tourism development (Moscardo, 2008).

There is need to comprehend different types of barriers in community capacity building that primarily relied upon degree of tourism activities. This topic is of critical significance during the time managing and organizing tourism activities.

This comprehension is very helpful to those individual, community and organizations that are directly related with tourism business and is concerned in tourism policy-making process. Moreover, government should also understand that the communities face several problems which can act as major roadblock in tourism development.

Addressing these barriers and issues in tourism development pose a challenge to both communities as well as government, and is helpful in framing policies and decision-making process. There are enough data and materials available that directly deal with the barriers of tourism development through local communities particularly in third world countries.

Power: Socio-political traditions: one of the major barriers that local communities in the development are the traditional beliefs of power-wielding governments that still exist in most of the third world countries. This form of government customarily nearly always precludes grassroots participation in tourism development (Sharma, 2004).

- Information accessibility: It is generally seen that lack of marketing and promoting abilities also hinder local business in developing and less developed countries. Local people do not have enough knowledge and skills about how to promote their sites and what are the possible ways to do the same.
- Limited access to advertising outlets, limited transportation services, and reservation system also contribute to lack of ability in some local business (Sharma, 2004).
- Lack of awareness: It is also observed that lack of awareness among people about the potential benefits and importance of tourism also prevent them to engage in these activities. Local people are often unaware about how increase in tourists in their area help them in improving their lifestyles by providing income and employment. It results in lack of participation by them in tourism activities and tourism decision making (Sharma, 2004).
- Economic problems: Lack of funding and limited allocation of budget is another limitation to local communities for achieving tourism development. Financial barriers at lower levels of administration result in high dependence on national government.
- It further results in lowering the involvement of local people in tourism decision making. In addition, it is also seen that there is lack of incentive and motivation to local people to participate in tourism development. Government should provide enough incentive to local people if it wants their participation in the tourism activities (Sharma, 2004).
- Lack of cooperation: It is always believed that for the success in tourism development, there is need to have proper coordination and cooperation among various stakeholders that are associated with the tourism development. It is often seen that in countries

where there is high level of cooperation, tourism development is more widespread (Sharma, 2004).

In this section, some of the common barriers and challenges that are perceived by Steven & Jennifer (2002) in local communities are explained. The important barriers to community capacity building are referred to as follows:

- Lack of understanding of policy process: It is important for local peoples, groups and community organizations to have good understanding about policy making process in order to decide whether they will become involved in trying to develop or change a policy and if, so, how to best go about it (Steven & Jennifer, 2002).
- Lack of community resources: In order for local communities to play an active role in the policy making, there is need to give enough resources to local community residents to have access to resources.
- These resources include appropriate financial support, government training and education programs to volunteers to support community development. It is generally seen that the major reason for failure is limited availability of these resources. Lack of funding and resources often have negative impact on the community's ability to effectively influence and develop tourism policy.
- Atkisson et al. (2003) believed tourism development at the community level has a number of limitations such as inadequate communication, lack of resources, and lack of communication and planning.
- Reliance on volunteers: Access to financial resources, which is essential in addressing various problems of a local community. As a consequence, limited resources availability and tourism expertise and skills from local community results in a depletion of potential future community leaders (Steven & Jennifer, 2002).
- Policy timeline restrictions: Restriction in policy making and implementation may also act as a major hindrance in capacity building. It is generally seen that policy timeline always create difficulties for communities that are looking forward to implement policies pertaining to a particular issue.

Lack of access to information: Lack of access to information is also major factor that why local people lack in putting their best in capacity building. Local people have indicated that they feel there is limited tourism information available to them. They also expressed concerns that the information that is available on government programs is also not easy obtain and interpret.

6.7. THE GROWTH PARADOX: CAN TOURISM EVER BE SUSTAINABLE?

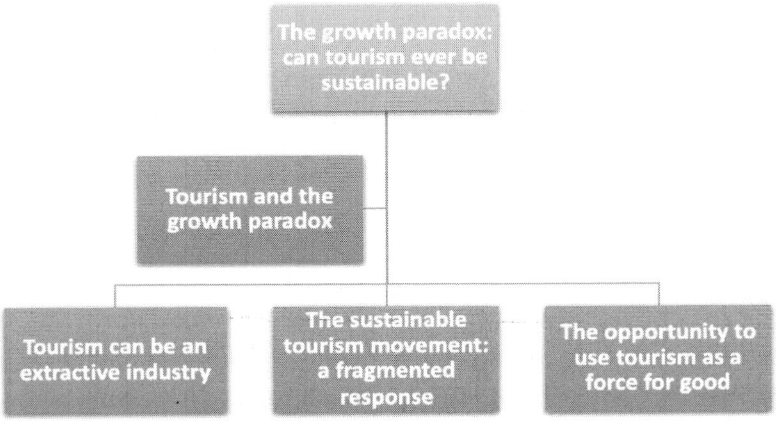

Figure 6.7. The growth paradox: can tourism ever be sustainable?

It is widely believed that tourism create jobs. In fact, a lot of jobs. It is worth considering that almost 10% of the population is employed in the tourism sector. When it is well managed, it can help in offering an incredible economic boost to host communities. However, because of all these advantages of tourism, almost every country in the world wishes to expand its base of tourism sector and want to enhance the total number of tourism arrivals. Hence, it is generally viewed that many travelers more likely to visit places and make purchases from travel companies based on their sustainable practices. Furthermore, there are several destinations that have their own set of policies and strategies that use tourism to tackle the challenges of poverty and conserve their natural and cultural heritage, for example, through the creation and management of National Parks and monuments. Yet, it was observed that almost 50% of World Heritage sites that are highlighted by UNESCO for their outstanding universal value, do not have tourism management organization that is interested with averting the negative impacts of tourism. On the other hand, if the increase in the tourism activity plays important role in providing employment to large number of people and help in economic boost to country, why there are an increasing number of destinations around the world proposing or implementing measures to limit or restrict it? Destinations such as Zion National Park, Machu Picchu, and Cinque Terre are restricting the number of tourists visiting their destination. Barcelona, the Seychelles, and Amsterdam are shortening large-scale development. Venice and Bhutan charge heavy fees in terms of taxes and visiting fees, while places such as Koh Tachai in the Similan National Park are forbidding visitation altogether. Moreover, the question raises: Why are not more destinations concentrating on yield per visitor instead of the number of international arrivals? Why are not more businesses that dependent on the tourism sector are investing in the destinations that they serve and the assets upon which their growth rely? And thus, why is not the industry as a whole putting effort in tackling these challenges and issues?

6.7.1. Tourism and the Growth Paradox

It is expected that global tourism will grow at a rapid rate as a larger number of aspiring travelers become more prosperous and their disposable income increases. It is anticipated that the sector will contribute heavily to the overall global economy in the next few years, increasing by an estimated 4% on average annually in the next 10 years. It is not surprising that majority of the destinations want to have a bigger piece of this pie and framed policies and strategies in place to attract a greater number of tourists and invest in the tourism activity. Taxpayers' money is spent on attracting visitors and the main area of focus is on the total number of tourists rather than the value and services provided to each visitor represents. Moreover, the net result is that the places served by ministries of tourism, convention and visitors bureau, and destination marketing organizations are often victims of their own success. Hence, travelers often come in large numbers and flock to the same locations, which result in difficulties such as overcrowding, cultural homogenization, increased stress on public infrastructure and services, and growing dissatisfaction from residents.

Nevertheless, local capacity constraints, ad hoc development, and leakage are issues as well. The latter one is the most prevailing issue: although tourism can be a great form of wealth distribution, it was found that only 5 to10% of the money that the tourism spends remains in the destinations they visit. These are some of the challenges that must be addressed to fully realize the benefits of tourism through an inclusive approach and collaboration between the private and public sectors as well as various host groups. Moreover, there is need on part of governments and travel industry to understand that the narrow focus on increasing numbers is a problem, which cause several issues and challenges that diminish the quality (and value) of the experience for travelers and visitors alike.

6.7.2. Tourism Can Be an Extractive Industry

One could argue that tourism cannot be sustainable, that sustainability is impossible. Negative effects on the environment are inherent to the industry, such as waste generation and emission from the greenhouse gases, that are currently challenging, if not impossible, to avoid. Tourism can also contribute to energy and water shortages, ecosystem degradation and degradation of water supplies, owing to ad hoc development and weak or poorly enforced environmental regulations. However, from clear cutting and destruction of mangroves to construction and excavations, land use changes that are associated with the tourism activity result in the creation of infrastructure and artificial landscapes. This gradually result in degrading the scenic value of natural landscapes and results in a change in the environment and climatic conditions, because of the certain factors such as pollution, erosion, and artificial boundaries. This further increases the chances of more erosion as well as may cause flooding and storm damage, may have a negative impact on wildlife migration and breeding patterns.

Figure 6.8. Tourism can be an extractive industry.

Source: Image by unsplash.com.

The tourism sector attracts increased investment in airlines, airports, railways, hotels, ports, and tourist attractions, especially in cultural and natural heritage sites, metropolitan areas, and other attractive landscapes and locations. Moreover, this short-term progress is good for economic development, but bad for residents living nearby when tourism development is poorly executed.

Local, historical and indigenous people are displaced or forced to relocate when new national parks or heritage sites are established. Prime properties are sold to real estate developers and local access can be restricted or lost altogether. Although it may result in creation of jobs, most of these jobs are either entry level or low paying with limited opportunities for upward mobility.

6.7.3. The Sustainable Tourism Movement: A Fragmented Response

Sustainable tourism is on track to go mainstream because it has the potential to combat the negative effects of tourism and enhance lives, offer visitors with authentic experiences and ensures preservation and protection of destination sites. But the sustainable tourism movement is as disintegrated as the industry itself. Moreover, today's pursuit of sustainable tourism is replete with individual consultants, and small organizations repeatedly combating for underfunded projects and small-scale, and scraps of success. It was found that most of the tourist management organizations have only one or two staff on payroll and even the largest NGOs have less than 25 employees. Hence, most of the focus in the public and private sector is placed on planning and development rather than changing operational or organizational practices, and follow-through on sustainability initiatives is confined because of "resource constraints."

This results in an intensive competition for the resources that are limited in nature and little meaningful collaboration. There was no solo organization that attained success at scale, and, despite good intentions, the cumulative impact is smaller than what is vital and achievable. Therefore, it is often seen that the term sustainable tourism has different meaning for different people in the industry. However, like ecotourism, sustainable tourism has gradually lost its importance as it is generally related to cursory efforts, which are very limited, rather than organization-wide strategies, commitments, and actions.

This is overwhelming considering that the business case for corporate responsibility as well as sustainability in tourism sector have been growing at an impressive rate every year. In fact, the UN declared 2017 the International Year of Sustainable Tourism for Development. Therefore, tourism is featured in three of the UN Sustainable Development Goals and all 17 goals can be advanced through sustainable tourism development. Moreover, this has given ascent to involve in activities that enhance overall environmental performance and social well-being. However, these practices also resulted in enhancing goodwill among customers, and employees, hefty savings, better employee motivation, lower absenteeism rate and higher retention rate and enhanced brand reputation.

Nevertheless, businesses that are directly related with such practices also benefit from attracting customers who care about front-page issues such as biodiversity loss, economic inequality, and climate change who commonly affiliate their spending with personal values when quality, price, and convenience are perceived to be comparable.

Still, there are very few tourism-related companies and destinations that have efficaciously incorporated sustainability into their operational DNA. Even there are only some companies that are investing in guard of the well-spring that their business depends upon — the destinations in which their business depends. Moreover, overall, the commitment of tourism sector to sustainable development is not impressive. It is understandable when viewing tourism from the perspective of most other industries that are majority of the times profit-driven and growth-oriented with a relatively short-sighted approach to planning and development. Hence, the main area of focus is on to get a return on investment to increase value of shareholder as early as possible, or, like many politicians, to appease voters and general public to foster the probability of re-election.

6.7.4. The Opportunity to Use Tourism as a Force for Good

It is essential for destination, companies, as well as for travelers to be more vigilant and held answerable for the activities done by them. They all individually play an imperative role in ensuring that the tourism lives up to its ability to enhance people's livelihoods and safeguard the environments upon which their lives relied upon. Therefore, it was found that each player is not fully seizing an opportunity to use tourism as a likelihood for environmental, social and economic advancement. Moreover, since there is no single stakeholder group that is only responsible for conserving a destination's natural and cultural heritage or generating economic outcomes or collaboration between sectors is crucial to channelize the transition required to keep healthy tourism destinations. However, there is a need to indulge in collaborative efforts by different tourist organization and business leaders to determine how to manage growing visitor numbers, address continuous evolving traveler expectations and shape and improve both visitor and host experiences. Hence, there is a significant opportunity for visitors, host communities, businesses, and local people to derive greater advantage from tourism.

On the other hand, the strategic use of public, private as well as community partnerships, for example, empowers host communities, provide supports to destination-level tourism strategies to ensure the protection of tourism assets and creates opportunities for business sectors and NGOs throughout the tourism value chain. Residents' well-being and the visitor experience are improved in the process. Cumulatively these efforts will help in ensuring science-based decision-making and help to minimize the negative effects of tourism activities. Furthermore, they can also help in reducing the commoditization of the beautiful and historic places upon which the entire tourism industry relies, while providing more opportunities for tourism and travelers businesses.

6.8. VARIOUS PROBLEMS IN PROVIDING TOURISM INFRASTRUCTURE

Figure 6.9. Problems in providing tourism infrastructure.

Source: Image by council on foreign relations.

It is usually perceived that there are many problems that the city planners face in the capacity building of tourism infrastructure. There are several areas that need to be improved in order to get the preferred result. However, in order to better tackle the prevailing challenges and difficulties involved, interviews with local people as well as tourism department groups and local stakeholder will be helpful. Therefore, some of the problems that are normally faced by government in tourism infrastructure development are discussed below.

6.8.1. The Absence Transportation Systems That Support Tourism

It is usually viewed that the local transportation system plays an imperative role in giving rise to tourism activities. Infrastructure facility is incompetent if it does not have proper transportation system. While building the policies related to infrastructure, it is often seen that proper transportation management is overlooked, which result in lack of transportation of facility for tourists. It therefore creates a bad image of the country where the transportation system is not advanced. This problem primarily lasts in developing and less developed countries. One of the reasons for lack of transportation is inadequate budget to offer transportation services.

Nevertheless, according to a survey, it was found that almost all countries have terminals for transportation but there is no proper scheduling for arrival and departure of vehicle. This limited transparency resulted in huge trouble for those whose tourists are relied upon public transport to go from one place to another. It is thus, worth noticing that transportation is one of the crucial parts of infrastructure. One of the main reasons that why there is lack of public transport vehicle on roads is because of the availability of high traffic on roads. Due to this reason, government of several countries avoid simply sidestep to run large number of vehicles on roads.

On the other hand, metro and subway services are best alternative of on road transport, but it requires heavy investment and is not reasonable for less developed countries. However, with surge in the total population, private vehicles on roads are unremittingly rising which is the major cause of high traffic. In order to provide smooth transport services, there is need to take proper actions either by developing more lanes road or metro services. Otherwise, it becomes very difficult to address this problem in future.

6.8.2. The Problems of Accessibility and Connectivity Between Destinations and Between Regions

One of the necessities of regional tourism growth is the formation of inter-regional connectivity. The connectivity prerequisite is the presence or the road network and rail network that connects between regions. Therefore, this is one of the most dominant issues in numerous countries. Providing connectivity to these routes needs huge investment, resulting in restricting many countries resists to develop these. Capacity building in infrastructure also demands acquiring of land, which in fact is additional cost for government. Moreover, it is well known that acquiring of land is not an easy job as it is the most cost expensive and face several criticisms from local people. It necessitates development of policies and strategies and engage in planning and decision making to tackle these issues and challenges.

6.9. CONCLUSION

At the end, it is concluded that tourism is one of the important sectors for economy. It provides a lot of opportunity to people as large number of people are employed in it. In addition, it is also significant source of gathering foreign reserves for any country. Hence, opportunities in tourism sector are limitless but it also comes with lot of challenges along with it. However, the majority of countries lack infrastructure needed to attract tourism. It is often the case of developing countries where there are lot of natural beauty, but they lack the basic infrastructure that can help in attracting tourism. Thus, an investment in infrastructure is worth considering, given the potential it offers in terms of revenue and employment generation.

REFERENCES

1. Ardito, L., Cerchione, R., Del Vecchio, P., & Raguseo, E., (2019). Big data in smart tourism: challenges, issues and opportunities. *Current Issues in Tourism*, *22*(15), 1805–1809.
2. Aref, & Gill, S., (2021). *Barriers of Community Capacity Building in Types of Tourism Activities*. [ebook] Available at: http://www.jofamericanscience.org/journals/am-sci/am0602/23_1206_Barriers_am0602.pdf (accessed 8 August 2021).
3. Aynalem, S., Birhanu, K., & Tesefay, S., (2021). *Employment Opportunities and Challenges in Tourism and Hospitality Sectors*. [online] Longdom.org. Available at: https://www.longdom.org/open-access/employment-opportunities-and-challenges-in-tourism-and-hos pitalitysectors-2167–0269–1000256.pdf (accessed 8 August 2021).
4. Bhola-Paul, H., (2016). *Tourism Challenges and the Opportunities for Sustainability: A Case Study of Grenada, Barbados, and Tobago*. [ebook] Available at: http://www.davidpublisher.org/Public/uploads/Contribute/55dc2c742d782.pdf (accessed 8 August 2021).
5. Budeanu, A., Miller, G., Moscardo, G., & Ooi, C., (2016). Sustainable tourism, progress, challenges and opportunities: an introduction. *Journal of Cleaner Production*, *111*, 285–294.
6. Deccan Herald. (2021). *Tourism and Jobs: A Better Future For All*. [online] Available at: https://www.deccanherald.com/opinion/panorama/tourism-and-jobs-better-future-718262.html (accessed 8 August 2021).
7. Dobour, N., (2021). *Problems and prospect of sustainable tourism development in the OIC countries: Ecotourism*. [ebook] https://www.sesric.org/files/article/127.pdf. Available at: https://www.sesric.org/files/article/127.pdf (accessed 8 August 2021).
8. Fonseca, F., (2012). *Challenges and Opportunities in the World of Tourism from the Point of View of Ecotourism*. [ebook] Available at: https://files.eric.ed.gov/fulltext/EJ1134232.pdf (accessed 8 August 2021).
9. FutureLearn. (2021). *Positive Economic Impacts*. [online] Available at: https://www.futurelearn.com/info/courses/the-impacts-of-tourism/0/steps/90650 (accessed 8 August 2021).

10. Igwe, I., Okeh, B., Duluora, E., Amalu, T., Oko, U., & Ehugbo, U., (2020). *Tourism industry effects on socio-cultural activities of host communities: evidence from Cross River State, Nigeria*. [online] link.springer.com. Available at: https://link.springer.com/article/10.1007/s10708–020–10151–1 (accessed 8 August 2021).
11. Jha, M., (2021). *Adventure Tourism: Opportunities and Challenges Galore*. [online] Traveltrendstoday.in. Available at: https://www.traveltrendstoday.in/news/india-tourism/item/7796-adventure-tourism-opportunities-and-challenges-galore (accessed 8 August 2021).
12. Lexology.com. (2017). *Hotels and Tourism – Challenges and Opportunities | Lexology*. [online] Available at: https://www.lexology.com/library/detail.aspx?g=25f08d72-ca0a-445a-9a3d-e3ca4f25ce49 (accessed 8 August 2021).
13. Mason, P., (2019). *Tourism Impacts, Planning and Management*. [ebook] Available at: https://www.skylineuniversity.ac.ae/pdf/tourism/Tourism%20Impacts.pdf (accessed 8 August 2021).
14. Mbaiwa, J., (2006). The Socio-cultural Impacts of Tourism Development in the Okavango Delta, Botswana. *Journal of Tourism and Cultural Change*, 2(3), pp. 163–186.
15. Medium. (2021). *Travel & Tourism Creates Livelihoods and Opportunities for Everyone in Society*. [online] Available at: https://worldtraveltourismcouncil.medium.com/travel-tourism-creates-livelihoods-and-opportunities-for-everyone-in-society-2acf5f4acec6 (accessed 8 August 2021).
16. Mullis, B., (2021). *The Growth Paradox: Can Tourism Ever Be Sustainable?* [online] World Economic Forum. Available at: https://www.weforum.org/agenda/2017/08/the-growth-paradox-can-tourism-ever-be-sustainable/ (accessed 8 August 2021).
17. Nune, S., (2017). *Tourism in India – Current Status, Opportunities and Challenges*. [online] Jagranjosh.com. Available at: https://www.jagranjosh.com/current-affairs/tourism-in-india-current-status-opportunities-and-challenges-1509369287–1 (accessed 8 August 2021).
18. Provia, K., Ronald, K., & Michelle, K., (2021). *Community Capacity Building, Local Community Involvement in Tourism Activities and Community Welfare in Uganda*. [online] Semanticscholar.org. Available at: https://www.semanticscholar.org/paper/Community-

Capacity-Building%2C-Local-Community-in-and-Provia-Ronald/53afaf227e5c3e2e15dba40697fd7a3ad1d06e04?p2df (accessed 8 August 2021).

19. Rangaiah, M., (2021). *What Are the Roles, Opportunities and Challenges Posed by Big Data in Tourism?* [online] Analyticssteps.com. Available at: https://www.analyticssteps.com/blogs/what-are-roles-opportunities-and-challenges-posed-big-data-tourism (accessed 8 August 2021).
20. Sharma, A., & Thapar, M., (2021). *Development of Tourism in the Third World Nations: A Comparative Analysis*. [ebook] Available at: https://www.witpress.com/Secure/elibrary/papers/ST16/ST16013FU1.pdf (accessed 8 August 2021).
21. SiteVisibility. (2019). *The Top Challenges Facing the Travel & Tourism Industry In 2019.* [online] Available at: https://www.sitevisibility.co.uk/blog/2019/03/08/top-challenges-facing-travel-industry-2019/ (accessed 8 August 2021).
22. Team, E., (2021). *Top 5 Challenges Confronting Tourism Travel Industry*. [online] Elite Asia Resource Center. Available at: https://resources.elitetranslations.asia/2017/08/29/challenges-confronting-travel-industry/ (accessed 8 August 2021).
23. UKEssays.com. (2021). *The Socio-Cultural Impact of Tourism Essay*. [online] Available at: https://www.ukessays.com/essays/tourism/the-socio-cultural-impact-of-tourism-tourism-essay.php (accessed 8 August 2021).

Chapter 7

Tourism Supply Chain Management

CONTENTS

In this chapter the tourism supply chain management has been discussed in detail. The various types of sourcing strategy of supply chain management in tourism have been discussed like capacity management, distribution strategy, integration strategy, customer service strategy, manufacturing strategy etc. has been discussed in detail in this chapter. Moreover, the components of tourism supply chain management like accommodation, transport, food and craft, destinations have also been discussed in detail in this chapter. Hence, the need of tourism supply chain management has also been discussed in detail in this chapter. The way tourism operations practices capture TSCM economic value has also been discussed in detail. The challenges in TSCM have finally also been discussed in this detail.

7.1. INTRODUCTION

Today's ever-changing business world today prevents companies from being able to survive on their own. The companies just trickle through as far as to make changes at any level of the supply chain. This strengthens the value of the topic of the supply chain network market and departs from the usual patterns in that it causes concern and interests competitors instead of different industry segments. Because of the important role that supplies chain play in manufacturing, the field of management has for the last two decades, study has also concentrated on studying how to expand and improve their ability. A lot of research has been done in supply chain management compared to other industry practices, though; perhaps because there is less of a need to expand this kind of management for the kind of operations used in tourism (Gengeswari et al., 2011; Rusko et al., 2009). Not only the representatives of the supply and demand, but even the successes of all of everybody else are important in order to achieving supply/demand fulfilment.

This fails in the service sector because products or services cannot be produced continuously and/service flaws are fixable, but by the time they meet consumers; it is met in the case of diverse commodities, because errors in manufacturing are rare and isolated quickly. In terms of market responsiveness, it is vital that all service providers on the supply side of goods or services perform their tasks in a timely, correct, and effective manner. Moreover, significant factor in the overall performance of the supply chain for buyers like wholesalers and suppliers are collaborating to maximize the growth of the suppliers' opportunity for revenue, both by working together and giving suppliers the freedom to conduct their business. the creation of a competitive advantage requires an analysis of both value and structure of the supplier-buyer relationship (Gules et al., 2009).

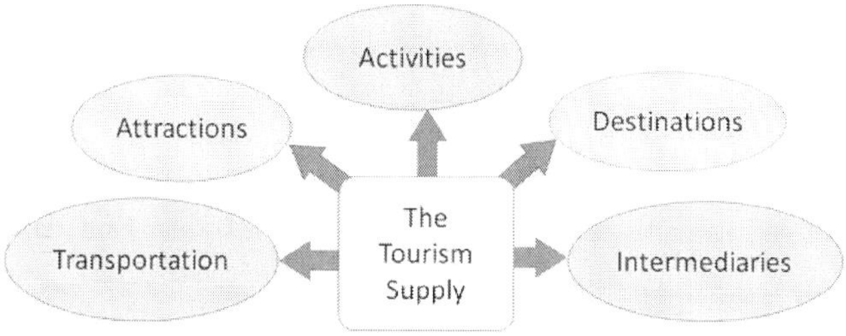

Figure 7.1. Tourism management supply.

Source: Image by Tutorialspoint.

The tourism supply chain research is mostly conducted upstream of the chain (Kozak et al., 2008; Zhang et al., 2009; Gengeswari et al., 2011; Tigu and Calaretu, 2013) and to measure the chain's performance (Yilmaz and Bititci, 2006; Tigu and Calaretu, 2013). A small portion of the studies on buyer-supplier relations in the tourism supply chain (Tektas and Kavak, 2010; Tao et al., 2009; Yildiz et al., 2010; Cakici and Cetinsoz, 2010). Although the results of this subject area have yet to be proven, there is a lack of literature. Even if suppliers' roles and impacts on the system must be given attention, they should not be disregarded.

7.2. TOURISM SUPPLY CHAIN MANAGEMENT

Tourism supply is a total includes all sorts of activities that go about maximizing the sharing of resources, cutting costs, and charging the right price, along with implementing value, all of which are applied to the final product, product, and also to final customers (Chen and Yi, 2010).

A variety of authors have written extensively about the definition of tourism supply chains in the literature in view of the distribution of a country's tourism supply-chain visibility. (Tapper and Font, 2004; Zhang et al., 2009) are basing their definition on country supply visibility on tourism chain distribution to the consumers. As in the case of Piboonrungroj and Disney (2009) and Chen and Yi (2010), there are several companies operating and many of them also cooperate on several projects. Therefore, many of the core functions of the tourism industry work to bring people in and to entertain them while most of the major participants in the chain consist of various accommodation, transportation, food and beverage, and shopping businesses, recreation companies, travel agencies and tour operators (Zhang et al., 2009; Tigu and Calaretu, 2013). Moreover, the second-level suppliers generating the first ring of the chain are the clothing, mobilities, fittings, appliances, water, and electricity etc. producers, to first-level suppliers providing lodging, catering, travel, shopping, and sightseeing services.

Nevertheless, when booking online, individuals have access to these tour operators via travel agencies or directly through the company. The tour operators who organize, plan, and offer to sell the tour package directly or through sub-hopper companies, in other words, take on the task of the delivery of tourism services distribution. Moreover, must of the people employed by the supporting tourism companies are people that stay away from home but offer basic services such as food and lodging to travelers and vacationers. The supply chain approach to tourism has grown up over time. Hence, due to the lack of extensive research, studies on the tourism supply chain tend to be few. However, the studies by Zhang et.al; (2009) look at the existing tourism supply chain have been examined in detail. They also described the key problems and the marks of tourism goods in the management of the supply chain. The obstacles to introduction of the tourist supply chain have been studied by Gengeswari et al. (2011). Finally, they further described these barriers to chain members' approaches (failure of trust between chain members, lack of coordination, refusal of integration among chain members), (ii) company structure (size, safety issues relating to documents, conflicts between structure), (iii) employees (incompetence of employees), and (iv) as a product and the structure of tourism.

Hong and Zailani (2011) have talked about the tourism supply chain, discussed the logistics structure, and its elements of the tourism supply chain. The researchers have described the implementation models of the service-supply chain into six activities and adjusts them to accommodate the tourism supply chain. The sustainability of the supply chain is a topic that Da Costa and Carvalho (2011) has covered extensively. They have evolved the supply chain by incorporating the value chain concept of tourism. They examined the tourism supply structure by identifying and describing all the various tourism activities and then delineating each one as a separate item in the value chain. Various reports on the management of the tourism supply chain are primarily on the Chain's Marketing Section (Kozak et al., 2008; Zhang et al., 2009; Adriana, 2009; Gengeswari and al., 2011; Tigu and

Calaretu, 2013) and to measure of performance within the chain (Yilmaz and Bititci, 2006; Tigu and Calaretu, 2013). However, a limitation of this new research on the subject is the limited scope of the studies that are incorporating buyer-supplier relationships.

7.2.1. The Buyer-Supplier Relationships of the Tourism Supply Chain Management

In general, the relationships between members of the tourism supply chain are competitive (price-focused) on one hand while cooperative (or trusting) on the other (Cusumano and Takeishi, 1991).

Figure 7.2. Tourism supply chain management.

Source: Image by unsplash.com.

Relationships that share this feature, such as customer-supplier and merchant and vendor, are nearly identical. Moreover, when placed in the West and Japan's complex relationships are considered, it is believed that the majority is of them are comparable to adversarial, but others are like those in North America and increasing closer to a relational model (Gules and Burgess, 1996). The communication and cooperation are essential to success in competitive-oriented relationship, particularly between suppliers and customers. For these reasons, the expansion of cooperation occurs mostly in compliance with requirements and is facilitated through the tight regulation of communication. With respect to this relationship, members of the same chain will compete against each other.

Nevertheless, the total number of suppliers and item specifications are shared, but pricing information is only provided to those who make the best deal. there is no impact on the creation or look of the product is done in terms of the things to order. This influences everything, including the production and the look of the goods and it is done without assistance from the vendors (Shapiro, 1985). The chain participants have chosen a cooperation-focused arrangement with each other based on a positive (or win-win) oriented mindset. In other words, the people on the chain that serve as corporate partners may be added as new possibilities. thus, making it easier for consumers to get the right details and understanding of their terms easily allows credibility to be established. By the revenue potential for the alliance, it will boost both parties' abilities to profitably cooperate.

In the long-term relationship-oriented approach, the number of suppliers is much less than the approach which is focused on attracting and expanding. This also describes where the buyer companies outline the specifications on which they plan to depend during the supplier selection process. Moreover, the supplier firms are evaluated and categorized by buyer firms and distinguished into various types of partnerships according to their different criteria. Therefore, buyer firms are assigned to certain kinds of suppliers in this, the collaboration-oriented arrangement, the cooperation-oriented environment is less stressful and competitive.

In this way, it leads to the emergence of new partnerships of trust and friendship with other people that previously do not have a cooperative relationship. In order to develop cooperation-oriented business partnerships, parties must have some partnership strength of their own. However, they must also have the capacity to keep their influence for a long time as per Spekman et al. (1998). Often known as continuous reduction of cost or increase of supply, firms' common goal is to improve overall cost of output while reducing the minimum number of defects (Shapiro, 1985).

However, few reports, just the relationship between buyers and suppliers within the tourism industries have been the subject of research (Tektas and Kavak, 2010; Yildiz et al., 2010; Cakici and Cetinsoz, 2010). In these works, Tektas and Kavak (2010) have investigated the influences of the hotel industry on the effect of tourism on its suppliers. In terms of how valuable they are to the hotels and how impactful they are to how trustworthy they are, what trustable they are, and what helpful they seem. Moreover, the executives have determined, the executives of the various hotel chains have found a positive and prominent effect on perceived importance, knowledge exchange and value.

On the other hand, Yildiz et al. (2010) have carried out a study to determine the primary criteria which businesses in the industry put forward when it comes to building new and improved relationships with their suppliers. Therefore, a detailed research study was conducted on hotels and concluded that they look at the creativity, the competitiveness, the added value to the corporation, their existing partnerships, the performance, how quick and adaptable they are, and how satisfied customers are in service delivery of their suppliers. Moreover, there is research underway to determine how open purchasing methods can impact consumers' tendency to purchase various products, on customers and brands (Cakici and Cetinsoz, 2010). Hence, the purchasing habits, as well as credibility, after sales support, capacity, reputation, and the importance of price have all found they have been strengthened by the findings of a study conducted.

7.3. SOURCING STRATEGY OF SUPPLY CHAIN MANAGEMENT IN TOURISM

The goal of supply chain management is to reduce redundancy in the supply chain and provide a better service to the consumer. In terms of non-value-added activities, such as the removal of excess, free flow of work, delays, and no longer maintaining inventory, reductions in costs can be seen. When customers get better treatment, they tend to buy from favorite vendors, and service providers. As a result, more preferred customers are won over and these suppliers realize a higher overall profit.

On the other hand, these kinds of regulations boost the competitiveness (income, efficiency). It is vital to do things well in a short timeframe to leave the customers with maximum quality or minimally inferior quality service. Moreover, the leading-edge firms, as a result of having shortened product life cycles, have an elevated level of rivalry, and anticipation from consumers, have found it necessary to improve their supply management capabilities. In most situations, the organization's scenarios, lower prices, and shorter lead times can only be accomplished by reducing supply chain management.

For what else, besides cost savings do reduce, will the supply chain strategy allow the company to enhance customer service. A successful warehouse management system must be capable of handling entire supply chains and logistics networks to allow companies to fulfil their goals and even greater potential for growth. Moreover, the hotel and travel supply chains all include several subcomponents, and facilities, as well as vehicles, meals, daily items, or retail stores. However, they also provide trash, warehousing, production, and catering and dining facilities, to name only a few. all the ties in the tourism supply chain contribute to the consistency of these holidays. Therefore, is one step in the tourism supply chain that consumers need to understand is buyers' behavior. Another step includes what the customer brings to the tourism business: on their own, including books, or things they would like to partake in on holiday. Moreover, a key role of any supply chain is to consider is obtaining the best materials, carrying out procurement and promotion, production, meeting the needs of the demand of the customers, and operation of the suppliers, and supporting the whole process. There are various types of tourism products and services that include hotels, transportation, destinations, event organizers, retail, travel agents, and airline companies, etc. In accordance with product design, SCM techniques may be updated.

The sourcing strategy are discussed below.

7.3.1. Make or Buy Decision

Owners could relate to direct delivery of services; also, the Thomas Cook method of doing things is more like an outsourcing a company that has established itself as a leader in vacation arranging. The implementation of a strategic partnership with Accenture included transferring the back-office work to Accenture. Just with the most simplistic change in the processes and cutting costs by $243.72 in a few simple models, they were able to expand its capacity and lower costs in 16 months.

7.3.2. Manufacturing Management

Manufacturing is restricted to specific industries including food and artisan souvenirs, but only one industry is permitted to expand. Therefore, at each point in the travel and tourism industry, services are done. as in many airports, such as transport systems, such as visa and immigration screening are provided to provide most travelers the highest satisfaction, in terms of wait times and comfort. It is to maintain and promote the health and safety of the visitors by ensuring the proper climate and protection.

Figure 7.3. Manufacturing management.

Source: Image by unsplash.com.

7.3.3. Capacity Management

Demand is one of the many challenges faced by the tourism industry. Example of the Muscat in Oman: the industry's Muscat during the festival, there is a great excess of hotel and airline availability at the Festival Season's demand. If this continues to increase, airlines attempt to satisfy this increased demand by means, they start to run more frequent and occasional flights at specially priced rates.

7.3.4. Distribution Strategy

In the tourism sector, delivery has a large effect on the development of new jobs and locations. If the channel strategy is done well, this is a must, as is the accurate and reliable handling of channel strategy between launch and closing.

The travel sector has three components: suppliers, middlemen, and consumers. Getting an understanding of the systems and structures where the parts of the industry work together is important to understand the Economics of tourism. Tourism is currently having four selling delivery forms: internet-based, brick-in-and-mortar store, offline, direct mail, catalogue, and magazine.

Figure 7.4. Distribution management.

Source: Image by unsplash.com.

7.3.5. Inventory Strategy

An analysis of local authorities has on tourists to examine the benefits and liabilities that a town/city/region brings to the tourist market helps with planning tourism goods. In addition to the list of attractions and events, the amount and breadth of diversity of experiences the visitors have is significant. people will avoid or skip locations where gaps in the product offering or inadequate services will have an adverse effect on the quality of visitors no place to stay, and no means of transportation would keep tourists from coming and make it less likely for them to remain on site.

It would be necessary to provide an inventory of everything which encourages the resort's visitors to remain (those things which retain them, such as events, accommodations, and transportation), which will be supported by surveys on visitor satisfaction be done until it is clear whether they are fulfilling visitor expectations. For the purposes of tourism, things such as attracting visitors and engaging them in various activities are the most important regions. as examples, visitors come to Switzerland to have fun and a service opportunity is something that helps them more than the tourists help the service providers. An important consideration of local tourism is the contribution of local governments to services such as libraries, museums, community centers, stadiums, and parks.

7.3.6. Customer Service Strategy

The customer satisfaction should be the main priority for the tour operator, which means that he must regularly fulfil each one or more of his customers' desires. when consumers do not know what to expect and get leads to inconsistent service Customer service involves offering accommodations, flight information, and making travel/tourist sites more interesting.

The Employees are critical to the success of this endeavor; they must help the customers to always be happy and satisfied. It is therefore important for organizations to have good processes in place in order to reach their targets. Dealing with clients properly would put the company in a worse position because team members are unable to do it correctly. Moreover, this rule of thumb works for both big customers and small ones: It costs about five times as much to acquire a big client as it does to keep existing ones.

Figure 7.5. Customer relationship management in tourism.

Source: Image by Pxhere.

7.3.7. Integration Strategy

To integrate the corporate institutions in the supply chain, Supply Chain strategy connects both. Tour operators perform both production and distribution functions in the tourism industry. Supply chain consumers can gain efficiency from use of a manufacturer or supplier's ability to use internal shared tools, consistent capabilities, and capacities to coordinate inventory delivery systems and fulfil orders in tandem with distributed networks to expedite the overall operation. the higher the amount of convergence in the supply chain, the more efficient and cost-effective it would be. One thing that service or tour operators must consider when designing a production is that they must get together successfully is the whole offering.

Figure 7.6. Sourcing strategy of supply chain management in tourism.

Source: Image by unsplash.com.

7.4. COMPONENTS OF TSCM

As changeable as well as variable service patterns are critical for tourism supply chain management, various tourisms categories must be handled in response to differing demand. notable facets of tourism include hotels, transportation, lodging, and crafts and destinations.

7.4.1. Accommodation

The hotels, restaurants, and entertainment venues also seem to be components of tourism products. When it comes to handling hotel facilities, the venue is one of the first things that is assessed at the start. In order to ensure convenience, it is open to the tourist's attractions, or beaches, the hotels must be placed nearby. Since the workers at the hotel should be dependable, proficient in modern methods, they should also be well educated and familiar with technology. Moreover, the program provided by the Hilton Real Estate Group handles their sourcing and inventory, which is called Birch Street. There is a significant need for more advancement in the Hotel Class and kind varieties in order to cater to all kinds of travelers. It is a commonly held belief that the restaurants and cafes are responsible for creating a picture of our community for customers. Therefore, it is important that the consumer have the ability and ease to use the product's features. There should be no lingering contaminants or residual hazards in the environment. Moreover, if the people staying in the hotel should be able to use the qualities of a quality-rated business hotel, such as comfortable beds, meeting halls, shops, cafeterias, pool, and taxis, the resort should give them some financial benefit in the form of free transportation. Hence, while delivering the product itself is important, an equally important service is the personal attention, quality of service provided to customers.

Carla Gold GM Radisson hotel, Cleveland, claims that the Radisson Hotel in Cleveland has expanded recently by over 25%. It is important for a company to be aware of its future customer demand because a strong supply chain begins with knowledge of the business. Moreover, effective inventory management is necessary in order to minimize the cost.

7.4.2. Transport

All the elements of the tourism package must be designed to make the travelers happy when they are traveling and must be conducive to their travel pleasure and experience when they arrive. Any of the airports are tourism landmarks. However, it is certainly not true that any of them double as tourist destinations are closely related to the industry. Most other forms of transportation use of public resources, particularly roads and railways, become much more important.

Figure 7.7. Components of TSCM: transportation.

Source: Image by unsplash.com.

For most tourists preferring to use their own buses, the tour operator would not have to provide facilities for the group. As this brings things to a new destination, this makes it critical that the transportation options are well-connected. Moreover, attractive, comfortable, safe, and affordable facilities should be available to tourists who want to maximize their vacation time. Since there are no alternative means of transport, the railroads and highways must be expanded to handle the expected increase in tourist traffic. Generally, you can keep a steady routine and pace of operation. The number of travelers to a country would have depends on how difficult it is for tourists to get visas and how fast they can travel through immigration procedures. Moreover, international border control has been modernized at Muscat Airport to make the wait time to get through immigration significantly shorter. Travelers are being given expedited to fill out forms to prevent them from delaying the whole system; automated passport control stations have been built to help speed up the process. Overall, Hong Kong International Airport was ranked the best airport in the world by Skytrax in 2017 for the third year in a row.

In order to encourage customers to feel like they are getting their money's worth by ensuring that the transport workers follow guidelines Seating should be appropriate to the task being performed. The pilots, flaggers, and other window workers must not harassers should be misbehaving. Moreover, the cost is irrelevant. Therefore, everything must be done to guarantee the integrity of the protection principles out of which it cost. It is therefore very important to note that under-sea travel season is a substantial bottleneck on the route to India, and during peak periods of the year this bottleneck hinders passenger movement. Charters help to expand the overall potential of the airline but are particularly helpful in the introduction of new destinations. Hence, attention must be paid to improving facilities for moving-stage facilities is necessary, particularly in tourist destinations.

Nevertheless, U.S., & international visitors ought to feel an attachment to trains for them to catch on. That is a generous way to describe it "palace on wheels" Caring for the coastline and inland areas are also equally essential. This opportunity must be thoroughly used by making cruises on the Kerala's coastline and the Sundarbans.

7.4.3. Food and Crafts

There are quite a few tours which focus on catering to locals and cultural features in addition to nature and wildlife, especially if they include accommodations, often using local foods and handicrafts. Food and craft products produced at lower costs of excellent quality will provide the local population with significant profit margins, so that a protocol is established for distribution and volume and consistent specifications are observed. Moreover, The Food Supply in the hospitality organizations and establishments will pose significant challenges because the highest priorities are having high-quality, reliable food at an affordable price and consistent prices, as well as adequate quantities.

Moreover, a dedication to sourcing initiative includes education and tools and equipment investment to ensure supply and distribution, including storage and distribution infrastructure to handle a higher-quality commodity as well as increased manufacturing and delivery network development of a robustness and distribution flow to get enough goods from various local suppliers into hotels to satisfy requirements. Hence, a tour operator has seldom if ever attempts to provide supply chain programs for food and crafts but attempts to get local vendors to buy from the same area to help in producing and selling food and other items nearby is widespread. Most small hotels depend on vendors to a greater extent, but some hotels with resources have programs to increase the levels of local supply they expect to use have introduced strategies to support local companies.

In a larger scale, these procedures are supported by executive chefs who encourage all-fostering and leaders who champion local procurement are essential for overall growth. When purchasing locally, it is important to take into consideration both financial contributions and job preservation. Local purchasing and manufacturing can both add money to the local economy and support local job retention. For example, the famous honeymoon destination of Hana in India is located near the mouth of the sacred Ganges River. It is possible to purchase anything one might dream of at the Humpoliah – lamps, jhumkis, clothes, armadillos, chicken curry, wooden puppets, various spices, locally-from-grown flowers, traditional healing techniques, musical instruments, and animal figurines. However, many are interested in the rural atmosphere as well as a leading too high ratings of tranquil tranquilly. Besides exploring the country's many shopping to your heart's content, you can also watch exciting and diverse folk dances and numerous dishes and delicacies from the subcontinent.

7.4.4. Destinations

As relates to the supply of infrastructure and transportation, the destination additionally provides support and serves the ultimate purpose of supporting the tourism industry. In order to enhance the sustainability, community infrastructure, environmentally and culturally conscious tourism organizations look to target methods such as reduced solid waste collection for residents, better access to clean air and more alternative modes of transportation, and a better understanding of waste disposal. It thus allows the community to take advantage of the good times and prepare for the bad times.

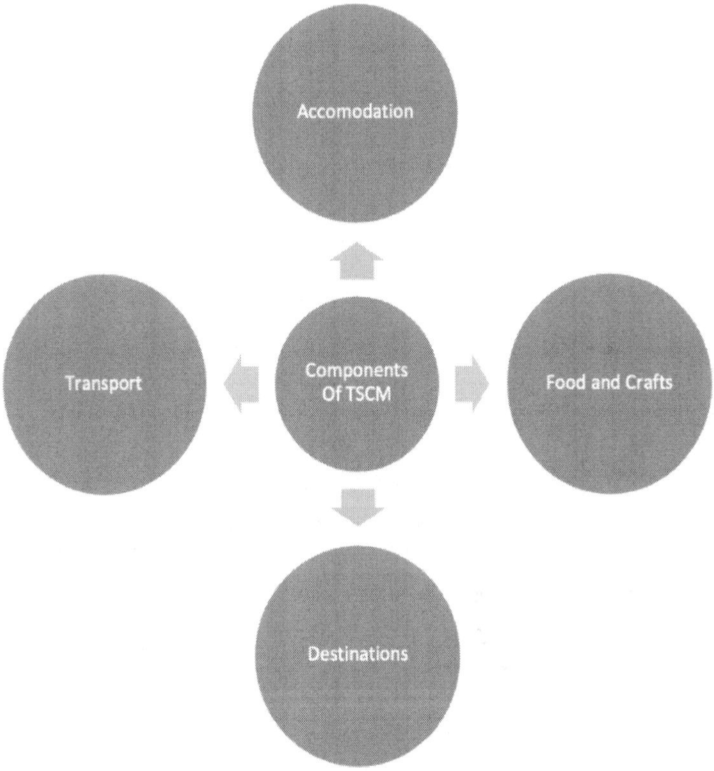

Figure 7.8. Components of tourism supply chain management.

The various efforts to increase tourism's network connections include promoting links between tourist destinations and tourism, and to increase the variety of experiences offered to tourists and enable them to sample the different parts of the world, making the distribution of economic benefits wider. Moreover, these initiatives help bolster tourism in locations, such as setting up business loans and investment funds that enable them to substitute local for imports, as well as investing in a local product manufacturing in order to assist in the expansion of the sector, while others focus on diversifying its scope to bring about jobs for the people who lack the skills to be used in a particular area of specialization. However, some destinations have developed business impact and environmental quality or visitation program initiative programs that rely on measurements, such as water quality and in the sustainability metrics and others are still maintain species diversity and markets. Adaptation will then follow on the results achieved from marketing, brands, products, and the same can be done about the control and optimization. Therefore, more than two million visitors come to see the Taj Mahal on the Seven Wonders of the World tour every year, and there are an additional 200,000 people who come to just look at it on the Taj tour. Hence, people living and working near tourism destinations will not be permitted to drive cars or use a public transportation vehicle.

7.5. THE NEED OF TOURISM SUPPLY CHAIN MANAGEMENT

All in the tourism industry considers supply chains to be important, including hotels, tour operators, suppliers, and travelers. Tourism industries can be served if the manufacturers of goods and services are made to co-operate with each other. As per Roy and Mamun (2016), the supply chain consists of all various vendors and plays a role in contributing to the experience of vacationers, so suppliers in the sector should be made to work together for the overall co-operation of tourism. Moreover, because that is the case, the main goal of a company is to earn the customer's appreciation and profit. These two values are always in direct opposition. Hence, Intojunyomg et al. (2016) also remarked that the use of a tourism supply chain management strategy (TSCM) enables effective operations of the tourism supply chain in a particular destination for the needs of tourists tailored to a specific business segment.

Overall, organizations are more and frequently endeavoring to assemble the framework of their supply chains, particularly regarding the weighty responsibility they place on their suppliers as per Roberts-Lombard (2009).

The tourism is simply a service that happens in a location because the essence of the experience is not in going, but in getting to the destination Henama (2018). The sheer scale of the service delivery of the industry enhances the risk associated with tourism. Moreover, the supply chain management's critical role in shielding businesses from difficulties that diminish production and profitability is observed by Simba et al. (2017). Hence, deciding on whether your company can be effective depends on whether you are able to provide suppliers with an adequate level of resiliency.

Without adhering to some supply schedules, marketing operations can result in a dynamic relationship between supply and demand, which can complicate things greatly, if not even bring about, erratic results. In the same way as supply chains in all other industries, such as agriculture and manufacturing, can contribute to financial success, the tourist industry and supply chain are able to have long-term benefits with a focus on activities in their supply relationships. The key distinction between tourism supply chains and those of other industries is that consumers visit the product, while the product itself has a significant service component: the latter brings plenty of people into play now of purchase (Tapper & Font, 2004). Opportunities exist for companies are prompting the need to achieve higher productivity, greater efficiency and responsiveness at the same time pressuring them to slash costs (Roberts-Lombard, 2009).

Thoo et al. (2017) refers to competitiveness, as well as to an environment that favors rapid market expansion which strong consumer satisfaction and pushes firms to excel in terms of their efficiency and low prices. Also, they believe supply chain integration (SCI) helps a business to build and hold on to its competitive edge through the ability to maintain constant, non-substantiable, and relatively impermanent path dependency. Moreover, when working with supply chains, problems have been encountered, problems have been found with a diverse set of criteria. While tourist businesses still seem to be on any promotional brochure, it is list of the pros, successful organizations have the following characteristics: monetary success, consumer loyalty, and sustainable activity (Fawcett, Ellram & Ogden, 2007).

More to the point, on that, having to strike a balance between supply and demand while not losing consistency can be a difficulty. Another important factor is being held in check in the business: transparency. Therefore, in order to be profitable, a corporation or organization must have a balance sheet which incorporates the elements of profitability, sustainability, and social responsibility. Moreover, stakeholder management includes considering the role of society, the environment, and the economy, but is especially concerned with the welfare of all stakeholders.

The tourism supply chain encompasses various parts, including housing, logistics, distribution, clean water systems, trash disposal, food supply chains, and vehicles, as well as the various other resources such as handicraft, packaging, public utilities, and sanitation in destination locations (Tapper & Font, 2004). In order to handle volume by increasing or decreasing market demand over time for various components or services in the supply chain to keep prices at levels where the company is either optimal or sustainable for the market price points (Song, 2011; Zhang, Song & Huang, 2009). Therefore, the method of (Song, 2011) places emphasis on managing the tourism supply chain and encouraging alignment and coherence, along with meeting demand at each stage in the tourism supply chain, in the work of Zhang, Song, and Huang (2009).

Nevertheless, an obvious difference between tourism supply chain and general supply chains is that presents a greater service aspect as products are supplied when visitors fly. It must therefore be bought and the merchandise is purchased (Tapper & Font, 2004). A supply chain planning and administration procedures which be used to contribute towards better end customer satisfaction by strategies in consideration of addition of the elements that go beyond mere infrastructure, which is as advocated by Fawcett et al. (2007). Moreover, despite the TSCM's research being in its infancy, they point out that there is no extensive research on the subject, which is an observation consistent with Piboonrungroj and Disney (2009). In addition, the article published by Zhang et al. (2009) found that while the early stages of TSCM study have shown to be observational, they remain largely hypothetical at this time.

On the other hand, the tourism supply chain studies use, or the works of Zhang et al. (2009) are often hailed as bearers because of their advances in this field. Their research is seen as a breakthrough in Piboonrungroj and Disney's research on tourism supply chain management (2009). This include Naude et al. (2013) papers on the management of supplier ties as it is a completely new concept for the South African procurement industry. Moreover, a few papers that pertain to supply chain management in Africa have been published, but few, if any, address Africa's companies. Hence, an investigation was carried out by the Naude and Derera (2015) about hiking to the summit of Kilimanjaro through the Machame trail, specifically about logistics.

However, many publications cover different aspects of African cultures, customs, beliefs, and experiences. In line with the topics, the emerging consensus would seem that tourism should include the management of supply chain, in view of the increasing vision of the need for a chain analysis of tourism in Buhalis (2001) (2003a), (2003b), Van der Duim and Caalders (2013), Song et al. (2012). In addition to the demands for more and requirements to move across the global supply chain are presenting new challenges for research and development, the supply chain has to move into a position in which it may perform functions that go beyond its original leading status. In view of the multifaceted and multi-level existence, TSC performance measurement is important.

Maybe it will be critical to address the position of supply chain. Supply chains allow new businesses to enter and set up shop, but at the same time restrict the development of new businesses. This type of growth is particularly crucial when it comes to bringing locals into the tourism supply when you are attempting to generate a livelihood for people who can live simply but still be wealthy and grateful host. Therefore, the growth in the benefits the disadvantaged community in that their small enterprises are used as suppliers, and employment opportunities that usually go unused for them are used to meet it. Hence, the tourism industry does not help thrive the target region until it incorporates the pro-poor values into the tourism industry.

7.5.1. The Way Tourism Operations Practices Capture TSCM Economic Value

As per Sanyanunthana et al., (2012) the organization of tourism industry supply chains is an important division because it promotes the greatest productivity in terms of expense, time, and consumer satisfaction. Further, they said that the tourism sector should strive to provide products and services that enrich or enhance the visitor's experience. In order to accomplish these objectives, acquire a higher logistics network and transportation planning network, as well as to secure a cost advantage over competitors and establish a value-added goal, decrease operating costs, and create a competitive position, are essential. Moreover, Sanyanunthana et al. (2012) stated that in order to meet the level of completeness in terms of efficiency evaluation and to cover the efficiency evaluation of the supply chain, policy, and target of increasing value added, and therefore to foster competitiveness among tourism networking collaborations across all sectors; government, corporate, community, international, and regional.

On the other hand, Naude et al., (2013) state that company organizations use supplier relationship management (SRM) in the public and private sectors. When dealing with complex transactions, they may want to use some tools to help ensure successful completion of the transaction: Necessarily, to stay in business in the long term, it is imperative that the development and establishment of long and value-adding supplier relationships be created and established. Hence, because all the things that make a relationship worthwhile have a positive effect on its other elements, such a relationship must be built on cooperation, confidence, and loyalty (Roberts-Lombard, 2009).

A lot of things are to be taken into consideration when forming and keeping a significant supplier partnership (Parvatiar & Sheth, 2000). These include.

- being sure that the arrangement is sound (is dependable) in terms of meeting deadlines as well as flexibility to respond in a timely manner.
- their ability to supply There are several different ways for the business and tourism sector to fulfil its goal; however, only those with other stakeholders which expand the goals of both sectors and foster beneficial ties expand their influence in the overall economy.

It is recognized by Stevenson (2009) as an important for maintaining a competitive edge. Only good partnerships with vendors are seen as a must for maintaining one. The private and public sector are also reliant on supply management and Naude (2013) states that company owners, alike, are of the opinion that both need a supplier partnership. The consumers insofar as there are financial and competitive goals to be met by the service companies, engagement with customers is important. Additionally, Sigala (2014) suggests that this could be a motivation for the organization to practice sustainable TSCM. One way of thinking about the tourism is the supply side. Suppliers in the tourism industry may think they can cooperate with customers in order to meet the demand; meanwhile, customers may want to see what other suppliers may do in terms of comparison, particularly when competitors come as supply is difficult (Song, Yang & Huang, 2009, Huang, Chen, Song & Zhang, 2010).

It is therefore important to manage consumer supply and demand, whether it be by customer participation, service activities, or even mobile apps. Moreover, the Supply chain management is geared to best meet the country and fulfil the needs of governments and/supply is procured equitably, equally, and to prevent monopolies and high prices while still increasing competition and effectiveness (Naude et al, 2013). Hence, based on tourism, Sakhuja and Jain (2012) propose a different service supply chain (information management, service demand, service customer, supplier, key operation points) processes that can be performed by working in tandem, and thus allow him to achieve this goal (deliver better service to customers, deliver on their expectations, etc.

Thus, by helping Sakhuja and customer as well as his clients, they provide better service, delivering on their customers' expectations, and satisfying expectations, J comes up with his goal, which calls for various operational elements will need to be kept together in an efficiently run supply chain (operations) with the services (demand, customer, supply, customer, and service provider, and service intersection, service, customer, provider, and supplier). Moreover, in order to ensure that our customers have the service they need, we intend to extend the delivery capabilities of the Transportation Network to various supply chain (Sakhuja and Jain, 2012). They contend that the general framework works in every service industry, giving activity managers the direction to perform their service operations systematically and on the other side effects to reach strategic goals are valuable for carrying out their company's objectives. we also Integration and interdependence

With regards to managing tourism supply chains, some proponents emphasize the value of efficient incorporation and networking of tourism organizations while others emphasize supply and delivery of goods and facilities and items to support tourism at a particular location. The only possible solution to the sector structure in a cohesive effort to include government and community participation in the expansion of tourism is to assume a third-sector leadership role in its form.

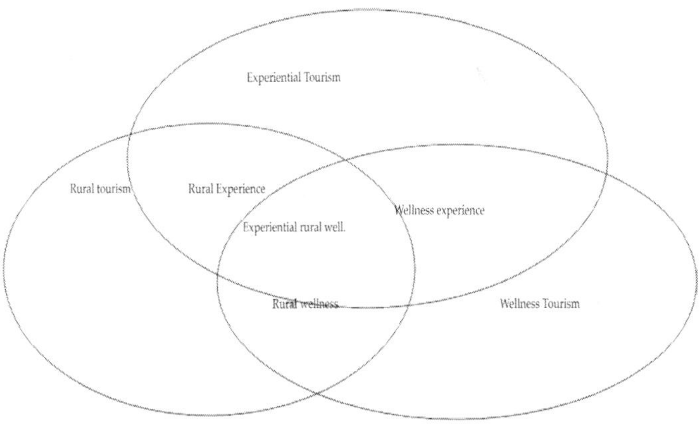

Figure 7.9. Tourism supply chain effect on local development.

Hence a public sector first, private second because there are already proposals for the supply management of the tourism industry. Finally there is an argument that making this one before the horses rather than trying to think about the different stakeholders who will be affected would be a kind of insanity. This kind of development is less concerned with outcomes than with achieving fair footing among stakeholders, as a first, and after that, attention will be given to any ways to improve the processes in place, especially the tourism supply chain. Therefore, comparative advantage cannot be ensured without working together and collaborating with others.

7.6. CHALLENGES IN TSCM

There are many challenges facing the tourism supply chain. In general, academics note that fragmentation in the tourism sector is an obstacle to sector development and thus a threat. Adversarial relationships are likened (though, unfortunately, within the larger society, it refers to non-native groups and does not always follow the same lines of communication) (Zhang, Song & Huang, 2009).

For example, the government is absolutely not unaware of this massive obstacle to the negative trends under which business enterprises and groups are exhibiting ever-increasing tendencies to divide, the markets, hence the NTSS understands that this as a crucial issue (National Tourism Sector Strategy, 2011). In contrast, the lack of a well-established tourism industry fragmentary in a region, creating various kinds of tourism opportunities for many competitors, which enables an efficient number of businesses to participate in the marketplace.

The key differences among academics seem to be in their opinions on the aims of tourism supply chains, but on the other grounds, there being little disagreement on that they have an interest in supply chains.' Both Tapper and Font (2004) identify six important facets of the tourism supply chain, and note that these elements: The accommodations, transportation, ground-handling services, events, and food and craft may all be components of a tour. Beyond that, Font, Tapper, Cochrane, and Cochrane (2006) state that three steps are required to make for a mutually successful supply chain between companies and their suppliers, they encourage implementing and sustaining supply chain policy. Along with that, they advocate for integrating sustainability criteria into the supply chain when prioritizing them.

While Font is not explicitly in disagreement with the primary issues addressed in the study of Tapper and colleagues (2004), there is significant discussion about seven critical supply chain topics, including supply management, development of parties within the supply chain, commodity, quality management, inventory management, transportation of goods, and information technology, all of which is included in Zhang et al. (2009). Moreover, the beliefs held in these other laboratories are likely to assist with the management of fragmentation and even on issues pertaining to sector cooperation. They contend that tourism services, including lodging, transportation, food and beverage services, and retail services, and sightseeing can also make up a big segment of the products generated by the Zhang et al. (2009). These studies are notable, not only in their variety, but also in their low productivity.

Some of the issues facing the supply chain in the tourism industry is lack of organization and inefficient regulations as barriers to creating a transformed and equitable economy (NTSS, 2017) Despite SCM requirements for promoting tourism, no change in accordance with any city. Moreover, with legislation in this area in various jurisdictions, there have been relatively few attempts to counteract unequal and uncompetitive procurement procedures, especially in the tourism has been implemented (Auditor General, 2016). Variability, inefficiency, and volatility introduced by consumer engagement may have a negative impact on long-term service production (Sigala, 2014).

Furthermore, Matolong (2015) claims that the following issues are obstacles to an effective SCM implementation: limited expertise, proficiency, and availability of capacity, all of which can be dealt with by officials who have the required skills and role in the plan. Since there are not enough tours and since there are not enough training opportunities in the travel and tourism sector, all of those that have expertise are not employed. Moreover, since the schooling system would not lead to travel industry employee advancement, many talents are underutilized (Tourism Sector Human Resource Development Strategy, Abridged Summary). The Insufficient planning and budget linkage: preparation, while these hinders SCM's progress, other unethical policies within the government prevent it from proceeding, and a long and wasteful regulatory process constrains it (Ambe and Badenhorst-Weiss, 2012). By putting in place sub-in apathetic supervision and unintelligent controls, they usually followed by unsympathetic and unregulated behavior, such as unscriptural behavior, including insufficient appropriations of funds, leads to additional problems including fluctuating costs (Auditor General 2014).In this instance, it is assumed that the struggle of the province's tourism sector is due to the absence of sufficient supply of chain accommodation/initiatives is due to an insufficient supply of new supply of accommodations in the tourism supply chain.

7.7. CONCLUSION

In its definition, Supply Chain Management (SCM) nature, the idea has spread far and includes anything from the source of products and function to the delivery to the client in between, and on-going supply process with everything within the handling functions and through an entire supply enterprise. Tourism involves these concerns in the tourism service chain, ranging from the procurement of raw goods and resources to delivery and distributing and, finally, customer loyalty will assess the success of SCM. The demand management is also an essential component of the SCM tourism. Moreover, a number of organizations include things such as wait time for travelers, how minimum, the way the service they provide is affected by the environment, and environmental safety while developing their supply chain management strategies.

The supply chain management is an asset in today's global marketplace that assists in lowering costs and making the most efficient use of resources while enhancing sustainability and leadership that delivers all of the things. The growth of the tourism is inevitably connected to both their tourism companies and the many tourism-oriented goods and retail businesses that are available. By looking at what customers want and providing those needs, businesses will gain a strategic edge in the travel services industry. Many partners are needed for effective product development in the tourism, especially for development in the supply chain since different operators exist in each link of the tourism's product cycle.

The different ongoing studies has shown that in recent years that the topic of tourism supply chain is also continues to grow and is, though a relatively minor portion of publications on the subject make the assertion the findings achieved as a result of the carried-out evaluations should be considered introductory to in-depth analyses of relevant literature. Moreover, selected problems already include travel and social and growth concerns that are quite generic, which means it is just important to start investigating if you are already dealing with issues related to tourism businesses and territories. However, it would be worthwhile for the researchers to go further in order to uncover the specific issues as well.

REFERENCES

1. Buyukkekli, A., Ozoglu, B., & Kemer, E., (2014). Buyer-Supplier Relationship in Tourism Supply Chains: A Research in Cappadocia Region in Turkey. *Journal of Applied Sciences, 14*(7), 660–668.
2. Manavhela, P., & Henama, U., (2018). The effects of tourism supply chain management practices on tourism operations in Pretoria, *South Africa*. [ebook] p.2. Available at: https://www.ajhtl.com/uploads/7/1/6/3/7163688/article_49_vol_7_2__2018.pdf (accessed 8 August 2021).
3. Tapper, R., & Font, X., n.d. Tourism *Supply Chains*. [ebook] p.2. Available at: https://icrtourism.com.au/wp-content/uploads/2012/09/TourismSupplyChains.pdf (accessed 8 August 2021).
4. Țigu, G., & Călărețu, B., (2013). *Supply Chain Management Performance in Tourism*. [ebook] Available at: https://www.econstor.eu/bitstream/10419/168779/1/aej-v15-i33-p103.pdf (accessed 8 August 2021).
5. UKEssays.com. (2017). Supply Chain Management in Tourism. [online] Available at: https://www.ukessays.com/essays/business/supply-chain-in-tourism-services.php#:~:text=SCM%20in%20tourism%20will%20start,and%20delivery%20of%20the%20services. (accessed 8 August 2021).
6. Zhang, X., n.d. *Tourism Supply Chain Management: A New Research Agenda*. [ebook] p.2. Available at: http://ira.lib.polyu.edu.hk/bitstream/10397/1129/1/18-TSCM.pdf (accessed 8 August 2021).

Chapter 8

New Business Environment and Trends in Tourism

CONTENTS

This chapter explains the significance of new business environment and trends in industry of tourism. This chapter also explains the role of digital technology in the sector of tourism. This chapter also sheds light on various new trends in tourism industry. This chapter addresses the importance of the marketing management trends in tourism and hospitality industry. It therefore provides highlights on the new microenvironment of tourism and hospitality industry. This chapter explains how to prepare tourism businesses for the digital future. Moreover, this chapter includes various sorts of sustainable tourism that have been playing significant role. This chapter emphasizes the several different technologies that helps in enabling the digital tourism economy along with the upcoming tourism trends.

8.1. INTRODUCTION

Tourism is the bundle of tangible products and intangible services that can help to bring most profound experience one can get. With respect to time, tourism has been changing from the ancient form of religious tourism to a few new forms. Tourism industry contributes to 9% of the global GDP and offers one from every 11 jobs. In addition, the number of tourists has doubled over the past 20 years and a propelling growth is expected in coming few years among all market segments. Moreover, tourism industry has an intense potential to grow and generate revenues. Let us see, the new shaping trends in tourism today.

Figure 8.1. New business environment and trends in tourism.

Source: Image by I luv nature.

8.2. DIGITAL TECHNOLOGY IN TOURISM

Today, tour operators and tourists, both are equipped with latest technology. It has empowered the tourism business managers and tourists to explore, discover and reach new places by facilitating online travel and accommodation bookings, and more.

Figure 8.2. Digital technology in tourism.

Source: Image by unsplash.com.

The tourists use various mobile apps as simple as a compass app, online booking apps, currency converter apps, world time apps, language translation apps, weather apps, google maps, and restaurant or accommodation locating services on their mobile device. The tourists can also add their own data to create overlay on the Google Map and explore all possibilities in visiting a location by using Google Maps API. Some apps help to plan the tour, find out cheap flights, local transport hubs, eating joints, and destination attractions. The apps enhance the tourist experience before, during, and after the tour.

Today, the large tour operators invest finances to create their own mobile apps. They are inclined to use as less papers as they can and prefer to send pdf documents of itinerary to their customers and insurance documents to insurance service providing clients. Therefore, they use most of the apps the tourists use; plus, they highly rely on some apps such as Trafalgar app, Passport to Tour app, mTrip app, which can connect them with the tourists on the trip, track the itinerary, and get on-the-fly information of the tour. Hence, since technology is making tour operators reach their customers and clients at the speed of light, the tour operators use it for promoting their business and various products, increasing their brand awareness, knowing tourists' preferences, and providing easy access to their products and services.

8.3. NEW TRENDS IN TOURISM

Tourism is an ever-changing industry. A few previously unknown or unnoted forms of tourism are establishing today.

8.3.1. Polar Tourism

Arctic and Antarctic polar regions have always attracted tourists. Polar tourism is a dynamically growing industry due to the efforts tour operators take to provide various attractions, destinations, and activities for their customers. Moreover, adventure tourists and common tourists who long for unique weather experience, solitude, and view of wildlife in its natural habitat opt for polar tourism.

8.3.2. Space Tourism

It includes orbital and suborbital rocket flights into the space. Riding into the space for recreation and unique experience was the idea behind this tourism. Until date, only very rich tourists paying very large sum of money could possibly realize the dream to see beyond the blue planet. In coming years, this extravagant tourism can be made available for common people too.

8.3.3. Dark Tourism

Dark tourism is the oldest form of tourism developed recently. Due to the fear and natural attraction to uncover mystery of death human beings always have, some tourists prefer to visit the destinations such as battlefields, places of violent homicides, or any places where large number of people lost their lives naturally or forcefully in the span of last 100 to 125 years. For example, tourists visit Pompeii to see the corpses of the victims of the volcanic disaster literally turned into plaster casts. Moreover, the volcano on Mt. Vesuvius had destroyed the ancient town of Pompeii. Though the disaster occurred and claimed thousands of lives long ago, the threat of more such volcanic eruptions still exists today. Hence, Pompeii is an archaeological and a dark tourism site.

Some more places the tourists visit for dark tourism are:

- **Chernobyl and Prypiat, Ukraine** – Tourists visit this place to see the ruins of the nuclear disaster that took place on 26 Apr 1986.
- **Ground Zero, USA** – It is The World Trade Center site attracting tourists since the 9/11 attacks that took place in 2001.
- **Auschwitz, Poland** – This dark site is famous for Nazi concentration camps with various rooms piled with thousands of pairs of glasses, shoes, and human hair. It displays the walls of the corridor exerting the lists of their victims' names and the dates of their death, and the house of the camp commandant.
- **Costa Concordia** – On the coasts of Tuscany, Italy; the site of this wrecked ship attracted tourists for around two years.

8.3.4. Voluntourism

It is taking volunteered vacation and touring for charity. People do not just go touring for recreation and fun but also to serve the community dealing with natural calamities. In case of flash floods or cyclones, the people in the affected areas need support. Some tourists voluntarily visit such places and extend their hands for help in whichever possible way they can. Therefore, tourists are traveling to care for orphans, for plantation, protecting wildlife, and similar other tasks.

8.3.5. Luxury Tourism

It mainly pertains to the rich business tourists, who strongly believe that time is of prime importance and they must pay to save time at any cost. Wealthy tourists are inclined to undergo unique experience such as staying at a private island, personal attention from the service providers and access to elite class attractions and amenities.

8.3.6. Culinary Tourism

The tourists who like to receive local culinary experience, like to tour for this purpose. They attend food festivals, food competitions, visit local farms, vineries, and cheese manufacturing companies, interact with local community or cooks for special culinary experience.

8.3.7. BRIC Tourism

The four major developing countries namely Brazil, Russia, India, and China have a great potential for driving global economy through hospitality and tourism industry. These countries are important for both inbound and outbound tourism. Global tour operators are adapting their tourism businesses to exploit the huge market these countries provide.

8.4. MARKETING MANAGEMENT TRENDS IN TOURISM AND HOSPITALITY INDUSTRY

The basis for this reflection and its consequent questioning lies on the unavoidable structural changes which characterize the awakening of the 21st century and which are determining factors for the mutation of the macro environment of societies in general and tourism business (Knowles, et al., 2004; Holjevac, 2003). It is possible to state that we are facing a new set of transformations which have made present everyday life more global, uncertain and dynamic (Tribe, 2010). Consequently, we notice the presence of a set of profound transformations in the socio-cultural matrix of today's societies, especially Western societies, where a new socio-demographic profile stands out. This profile is therefore characterized by a new tendency towards the decrease of birth rates, the increase of average life expectancy, changes in the concept of family, a growing urbanization and a unique and simultaneous coexistence of four different generations (Traditionalists, Baby Boomers, Generations Xers, Millennials). Therefore, these generations have distinctive socio-cultural characteristics due to the structural differences which have shaped their everyday life and arose unique challenges in what concerns communication and interpersonal relationships (Lancaster & Stillman, 2002).

Nevertheless the "Millennials," also known as "Generation Y," are the clients of the future, because "Millennial's access digital media on daily basis and have the ability to communicate with and purchase from suppliers anywhere in the world." (Mangold & Smith, 2012:141), and so it is vital to acknowledge the importance of such profile. Moreover, the exponential development and increased dependence of technology in the context of the present society place it as the central paradigm of the social and economic development and thus changing the limits of ability and means of production.

Figure 8.3. Marketing management trends in tourism and hospitality industry.

Source: Image by unsplash.com.

Considering its impact on the lifestyle and communication of the population, the internet is probably one of the most important elements of this revolution. It has deeply changed the notion of time, shortening reality and establishing new lines of spatial, time and management organization (Abrate et al., 2012). Therefore, on the economic level, there is the emergence of new markets, especially the binomial resulting from the interaction between emerging and developed economies (Yeoman et al., 2012). Moreover, this emergence is the result of the extension and intensification of globalization, which for the first time puts the concept of market, and consequently business competitiveness, on a universal level. However, the rise of new trade blocks such as BRIC (Brazil, Russia, India, and China) tends to alter the natural balance of the markets in which "new consumers Markets will contribute to transition economies (Central and Eastern Europe) and in developing economies" (Asia and South Asia) (Yeoman, 2008:25).

On the other hand, the financial markets crisis, namely of the subprime and sovereign debt, limited not just the immediate prospect of public and private investments, but jeopardized the entire paradigm of economic development, thus creating a climate of unparallel uncertainty and insecurity, especially within developed economies. Regarding the environment, sustainability is represented in the "green argument" as the vital assumption of the model of sustainable development and therefore tends to be a requirement, rather than an optional or differentiating matter. Moreover, the evidence of environmental limitations, namely the increasing impact of climate change, has turned this matter into a fundamental issue in the agenda for the 21st century, thus forcing social awareness and demanding new behaviors, attitudes and actions towards the environment (Kang et al., 2012). However, on the political level, the terrorism portrayed in the events of the 9/11 has become an unavoidable reality and the feelings of insecurity and volatility have altered people's lifestyle and encourage short-term perspectives rather than the traditional long-term perspective (Taylor & Chesworth, 2005).

There is also the distinctive rising of neoliberal ideals which lead to an increasingly more open market. The private market tends to develop thanks to a growing number of companies that, facing no barriers and resulting from such liberalization, tend to compete in different markets on a global level. Moreover, the progress in the neoliberal philosophies is also seen in other aspects such as the significant decrease in the state's actions regarding the "Welfare State" and consequent cuttings in social benefits like paid holidays and Social Security. Hence, the "flexibilization" of different legal frameworks, like the labor market, is also noteworthy.

8.5. THE NEW MICROENVIRONMENT OF TOURISM AND HOSPITALITY INDUSTRY

8.5.1. Tendencies of the Tourism Demand

Throughout times tourism market has registered transversal characteristics like the increase of its supply and demand, which is something evident when we analyze international tourism demand data. Therefore, according to the United Nation World Tourism Organization (UNWTO) this demand was around 25 million in 1950 and today it has reached 1.000 million tourists (UNWTO, 2013). On the other hand, also according to the UNWTO forecast for 2030, it is expected that international tourism increases in the next 10 years as much as in the period between 1950–2010. Therefore, we could face a tourism demand of about 1.8 million tourists in 2030 (UNWTO, 2011).

This is clearly an exponential increase of tourism demand resulting especially from the increasing development of the world economy, in particular emerging economies like Brazil, India and Dubai, and political openness of countries such as China and Russia. Moreover, this scenario of increased globalization of the tourism activity is clearly stimulated by the intense development of technology and reflects on the growing accessibility of new destinations especially due to developments in air transport. However, it should be noted that Europe remains the major destination of international tourism and the greatest producer and receptor of tourists, despite the percentage decrease in worldwide tourism activity. Western European countries, as well as Middle East and Asia, are major emerging destinations. And this makes competition among destinations a vital variable of the tourism market.

Given this reality tourism demand is likely to become disperse. The five main tourism destinations regarding international tourist arrivals held 71% of the market in 1950 whereas today (2009) this number decreased to 31%. In contrast destinations other than those on the Top 15 represent already 44% (and just 3% in 1950).

8.5.2. Tendencies of Tourism Supply

Given this growing globalization of tourism because of the circumstances, tourism supply has evolved in an attempt to answer today's challenges. Regarding aviation, development and consolidation of airline alliances (such as Star Alliance, One World, Sky Team) alongside the technological development of flight equipment, reflect the need for an increasingly global and efficient air transport. Therefore, the dispersion of the tourism activity on a worldwide level has exceeded the operating limits of traditional international legacy companies and forced them to develop in cooperation and integration processes. From simple interline agreements to codeshare, this reality has evolved to the currently dominant model of airline company integration: alliances.

However, air transport market has already shown signs that this is not the end of this integration process, as the recent mergers and acquisitions between airline companies like KLM-Air France, British Airways-Iberia and Continental-United have proven. Moreover, the dispersion of tourism activity towards new destinations, namely the Middle East, has fostered the development of air transport in this area (Emirates, Qatar Air, Etihad, etc.), a fact that is proven by the number of flights equipment scheduled to be delivered until (2013). As result, "the air transport's center of gravity is moving eastwards" (Airbus, 2012:41).

Nevertheless, the increasing relevance of airlines in the Middle East comes from the geographical positioning of their hubs, and especially relevant in long haul flights based in the hub & spoke model, as well as the important economic and financial advantages these companies have due to their shareholding structures (with available capital to invest) and access to strategic natural resources (low-cost fuels, which represent 1/3 of the entire cost structure of an airline) (Doganis, 2001). Therefore, alongside the cooperation of the legacy companies that have also witnessed the development of the low-cost market it has proven itself to be an alternative supply especially in medium haul flights and which has translated into an increase of the business competitiveness and decrease of airline companies' profits (especially legacy).

Due to this circumstance, as well as the current business context, airlines are presently facing a challenging scenario that is translated in a return on capital insufficient to afford the cost of capital (IATA, 2010). However, the phenomena of the concentration of the tourism supply associated to a philosophy of worldwide service is not limited only to the airline business. In the hotel business we also witness the growth and development of huge international hotel chains like Accor, Starwood, Marriott or Intercontinental. Such growth is both quantitative and qualitative. Moreover, if on the one hand the hotel chains have proven their commitment to the development of a worldwide supply (with recent and frequent openings in emerging destinations like the Middle East), on the other they have also developed their brand portfolio aiming even more to an increasingly specialized demand.

For example, the Accor group, has now four different levels that vary from budget to luxury and upscale (ex: Formula 1 -1 star-, Ibis -2 star-, Mercure -3 star, Novotel -4 star- and Sofitel -5 star) (Accor, 2011) and has recently launched the brand hotelF1, slightly different from the more traditional Formula1 because it is not just a budget hotel, but a design budget hotel. In travel agencies we notice a similar phenomenon. Thomas Cook and TUI, which clearly dominate the European market, have presently a business portfolio of over 300 brands. Its scope has widened, and they are now working far beyond their initial purpose of intermediaries and wholesale and retail businesses in the tourism activity.

This is also noteworthy in other tourism sectors such as the rent-a-car, like Avis, Sixt, Europcar or Hertz, and the cruise ships market owned by two major companies: Carnival Corporation and Royal Caribbean. Their wide brand portfolio is the result of the evolution of the cruise market through various mergers and acquisitions. Carnival Corporation owns brands such as Carnival, Princess Cruises, Costa, P&O, Holland America Line and Cunard. Such a universe of brands and services and wide range of concepts ensures the parent company a global positioning.

8.6. PREPARING TOURISM BUSINESSES FOR THE DIGITAL FUTURE

Digitalization is changing the way people live, work, and travel, and has opened new opportunities for tourism businesses to compete in global markets. This chapter examines the impact of digitalization on tourism, with a particular focus on SMEs. It therefore highlights the important role of government in creating the right framework conditions for the digital transformation of tourism business models and the wider tourism ecosystem and identifies several key policy considerations to foster digital technology uptake and use by tourism SMEs. Moreover, the evolution and application of digital technologies are profoundly changing the way people live, work, travel and do business, and in the process, they are transforming and reshaping tourism. The scope and uptake of digital technologies varies across countries, sectors, organizations and places.

However, the resulting opportunities and barriers create an uneven playing field, which is exacerbated by a growing gap between tech-driven and globally connected tourism businesses, and traditional micro and small businesses often characterized by low-tech business practices.

Figure 8.4. Preparing tourism businesses for the digital future.

Source: Image by unsplash.com.

Much attention to date has focused on digital marketing and e-commerce as a way of reaching new markets, engaging customers and building brand. However, while these technologies might build market access and awareness, increase connectivity and facilitate financial transactions, they are less effective in enhancing productivity or innovation in an increasingly competitive global marketplace. Therefore, productivity-enhancing technologies (e.g., cloud computing, data analytics, revenue management software) have generally received low uptake in tourism, while innovative technologies (e.g., augmented reality, geotagging) are generating, customizing and delivering in ever more novel ways, new visitor products, services and experiences (OECD, 2018c). Moreover, digital transformation is thus pushing tourism in new and often unpredictable directions. Digital technologies have important implications for tourism businesses of all sizes, for the structure and operation of tourism value chains and for the sector. Hence, facilitating and enabling digitalization in tourism is therefore a key policy challenge.

Nevertheless, digitalization is bringing unprecedented opportunities for tourism SMEs to access new markets, develop new tourism products and services, adopt new business models and processes, upgrade their position in global tourism value chains and integrate into digital ecosystems. It therefore brings significant potential benefits to SMEs. Moreover, it can help them to become more efficient, free up time and resources to focus on strategic tasks, and increase their capacity to develop new business models, enter new markets, or internationalize operations. However, SMEs are lagging in the digital transition (OECD, 2019d), and many small traditional tourism businesses are struggling to understand the opportunities and reap the benefits.

On the other hand, SMEs that do not invest in their digitalization will not survive, let alone thrive in the future. Moreover, destinations, businesses and the wider tourism sector will need to fully embrace these new technologies to remain competitive, and to take advantage of the innovation, productivity and value creation potential. Hence, policy makers have an important role to play to help tourism businesses of all sizes, including the more traditional and smallest firms, to engage with the digital revolution, and thrive in response to these paradigm-shifting technologies.

This chapter further discusses how digitalization is transforming tourism business models and processes, and the integration of tourism SMEs into global value chains and digital business eco-systems. The discussion is framed around three core key themes: how digitalization is reshaping the nature in which tourism businesses operate; emerging business models; and policy approaches to support digitalization of tourism SMEs. It is hence informed by responses to a survey of OECD Member and Partner countries, as well as wider OECD work on the Going Digital integrated policy framework which highlights a range of policy dimensions where governments should focus their efforts facilitate the right conditions to enhance digitalization.

8.6.1. Digital Trends Driving Change in Tourism

Digitalization is the process through which technology and data-driven management is transforming our social and economic systems and lives. The push to adopt digital technologies is driven by the convergence of advanced technologies and the increasing social and economic connectivity unfolding under globalization. Therefore, it has the potential to boost innovation, to generate economic and environmental efficiencies and increase productivity, including in the highly globalized tourism sector (OECD, 2017a). For example, research from Australia shows that using digital tools can save small businesses in general (defined as those with between 0 and 19 employees) 10 hours a week and can boost revenue by 27% (ANZ, 2018). Moreover, the capacity of tourism businesses of all sizes to evolve their business models, adopt digital technologies to effectively participate in global value ecosystems, and take up new ways of data-driven working, will shape productivity and social and economic wellbeing in the future (Andrews, Nicoletti and Timiliotis, 2018).

Digitalization leverages digital technologies and data to transform business models and practices, and value ecosystems. For example, the sharing economy has evolved in the last 10 years as a result of new platform technologies and business model innovation to create new value from hidden or unused assets. The value of the ride sharing sector in 2019 was estimated at USD 61 billion, while the value of the sharing accommodation sector is expected to reach USD 40 billion by 2022.

PwC has estimated that the value of the sharing economy will reach USD 335 billion by 2025. However, digital transformation is variable, and data is patchy about the uptake of different technologies and the barriers and opportunities experienced in different sectors, countries and organizations (PwC, 2016). Moreover, recent research undertaken by the European Commission found significant differences in the uptake of digital technologies in tourism across Europe. Nordic countries for example, exhibited higher uptake of digital technologies than those in eastern and southern Europe. However, the same research found that tourism SMEs lagged large enterprises. While basic e-marketing and e-commerce were widely adopted, advanced technologies such as data analytics, cloud computing and geotagging had received only limited uptake (Dredge et al., 2018). Therefore, with consumers increasingly using digital technologies to search, plan and book travel, it becomes increasingly important for tourism businesses to incorporate digital technologies and leverage advanced capabilities. Moreover, despite the uneven uptake of digital technologies by tourism SMEs, the digital transformation has, and will continue to have, a profound impact on tourism. The digital economy is transforming the process of communicating with tourists and marketing tourism services and opening new and highly creative ways of delivering tourism services and enhancing the visitor experience. Hence, it is changing the way work is organized and services delivered and presents opportunities to take advantage of digital advancements to handle transactions, capture and process information and data on tourism supply and demand, and improve and connect operations along tourism value chains and ecosystems.

On the other hand, according to the World Economic Forum (WEF, 2017), has estimated that in the decade to 2025, digitalization will create up to USD 305 billion of additional value for the tourism sector alone through increased profitability, while around USD 100 billion of value generated in the sector will transfer from traditional players to new digital competitors with innovative business models and value-producing capabilities. Moreover, the digital transformation is also forecast to generate benefits valued at USD 700 billion for customers and wider society, through reduced environmental footprint, improved safety and security, and cost and time savings for consumers.

On the demand side, this will be driven in part by the consumption habits of Millennials (born in the early-1980s to mid-1990s) and Generation Z (born in the late-1990s to early-2010s), who along with other emerging generations will comprise the bulk of domestic and international tourists by 2040 (OECD, 2018a). For example, Gen Z and the Millennials are digital natives and having grown up with quick and direct access to information enabled by digital technology, the way they use and what they expect from technology will continue to influence how tourism services are delivered. Moreover, trends include: increased use of online sources and mobile platforms to source information in the planning stage (e.g., websites, social media), combined with decreasing use of offline sources (e.g., visitor information centers, print media, hotel concierge); a propensity to stay online/connected in the destination to search and explore, share experiences and get updates in real time; and an increase in the use of e-commerce payment methods over the use of cash. Furthermore, Millennials and Gen Z have embraced sharing over ownership more than previous generations, and as such have higher levels of engagement in the sharing economy (accommodation sharing, ride sharing, currency swap and crowdsourcing).

On the other hand, while many of the challenges and opportunities for tourism SMEs are similar as those for SMEs more generally, tourism is a unique proposition for several reasons. For example, the tourism sector is highly fragmented and heterogeneous and covers a wide range of industries with many demonstrating a dual structure characterized by a very small group of large businesses combined with a large group of SME/micro-businesses. The sector is also 'information intensive,' which means many tourism services are ripe for digitalization. Moreover, understanding digital uptake by tourism SMEs is particularly relevant as around 85% of those enterprises with a major role in the delivery of tourism services in OECD countries are SMEs (e.g., accommodation and food services, travel agencies, tour operators), compared to roughly two-thirds for the wider economy.

Although they constitute most tourism businesses, SMEs and micro-companies face more difficulties to vertically integrate than larger companies (such as hotel chains in the accommodation sub-sector) and to reach potential customers. Moreover, another particularity of the tourism sector is that tourism enterprises operate in a global marketplace while delivering at the local level, as part of a unique tourism destination offer. The geographical distribution of businesses is limited only by the attractiveness and accessibility of destinations. Hence, digital businesses such as online travel agents and accommodation platforms, have transformed tourism by connecting tourism products and services with customers anywhere in the world in real time, as well as significantly increasing market visibility for tourism SMEs. As a result, many tourism SMEs are at least partly dependent on larger intermediaries, while at the same time being under pressure from increasing consumer demands for quality and efficiency (e.g., the best service possible at the lowest price possible).

In the case for many sectors, lifestyle and micro-enterprises in the tourism sector have a reputation for being focused on business survival, and being risk averse with limited appetite for innovation, technology pervades through most aspects of modern tourism businesses. Technological developments are progressively changing tourism value chains and the position of SMEs within them, enabling business models to evolve, and offering new ways for SMEs to collaborate and network with potential partners (e.g., to present seamless integrated visitor experiences). Moreover, tourism value chains have transformed into global value ecosystems as consumers can now have direct access to the businesses in charge of delivering the final tourism good, service or experience. Hence, informed by digitally derived data, tourism products and services are evolving toward hyper-personalization and customization (Skift and Adobe, 2018; Visa, 2017).

Nevertheless, automation and robotics are a key trend in tourism with applications developed for physical tourism businesses as well as online. For instance, many tasks that were once handled by humans now being taken over by robots or automated systems such as chatbots. Therefore, these have fallen into widespread use across the industry and are designed to help people find and book tours, transport and accommodation by asking a set of questions. Moreover, some sophisticated examples might include the robot "staff" used by some hotels to run the reception desk or even serve food and drinks. While robots have advanced in their abilities to provide products and services, industry, government and consumers have not entirely figured out how to integrate these into the economy (Ivanov and Webster, 2019). Hence, the development and adoption of new technologies is expected to continue at pace in the future, driven by the cumulative nature and exponential rate of technological change, the convergence of technologies into new combinations, dramatic reductions in costs, the emergence of new digital business models and declining entry costs (UNCTAD, 2018).

However, previous OECD work on enabling technologies shaping the future of tourism highlighted the need to develop a better understanding of the challenges and opportunities arising from these technological advancements to inform the development of appropriate policy responses (OECD, 2018a; OECD, 2019a). Therefore, the shift to a digital economy offers opportunities for tourism enterprises of all sizes including access new markets and bringing new tourism services to consumers globally, and improving competitiveness, performance and productivity. For example, for SMEs, it can improve access to market intelligence, enable businesses to achieve scale without mass, and facilitate access to global markets and knowledge networks at relatively low cost (OECD, 2017b). On the other hand, the digitalization of SMEs has been identified as a particular challenge in relation to productivity in micro and small business that are often resource-constrained (European Commission, 2017; OECD, 2019a). OECD data shows that while the gaps in the uptake of digital technologies between large and small firms have narrowed in most countries in terms of simple connectivity and web presence, these gaps remain more important for more advanced technologies.

8.7. SUSTAINABLE TOURISM: MOST IMPORTANT TRENDS

As academics specializing in the travel, tourism and hospitality (TTH) industry, we have trained and advised dozens of companies in their journey towards sustainable businesses. There is so much more sustainability than just the standard greenwashing statements. Moreover, TTH firms are accountable and must take the changes taking place seriously. From Greta Thunberg´s movement to new priorities in the World Economic Forum´s agenda or the ever-growing list of world moguls taking on a leading role in the fight against climate change. For that reason, we have identified 10 sustainable tourism trends that will influence business strategies in 2020.

Figure 8.5. Most important trends in sustainable tourism.

Source: Image by unsplash.com.

Each trend is an opportunity for TTH firms to provide services that align with consumer values, the Sustainability Development Goals agreed by the Member States of the United Nations as part of the Agenda for Sustainable Development in 2030, and the triple-bottom line accounting framework to evaluate business value in a broader perspective that includes social, environmental and financial outcomes.

8.7.1. The Time for Isolated Green Initiatives Is Over

Greenwashing is ever less acceptable. Current challenges demand a comprehensive sustainable business model. One that places sustainability at the core of your corporate strategy, business practices and operations.

8.7.2. Carbon Emission Compliance Is No Longer an Option

TTH firms need to have a greenhouse gas inventory for their companies. Still today not many companies have one.

8.7.3. Implementing an Ambitious Emission Plan Pays Off

Companies are accountable for reducing their emissions. Some can try a modest carbon-neutral way. A better objective is to reduce overall emissions by 5% each year. Moreover, leading TTH companies should go all the way down the road just like Microsoft recently did. Hence, when it stated that by 2030 "Microsoft will be carbon negative, and by 2050 Microsoft will remove from the environment all the carbon the company has emitted either directly or by electrical consumption since it was founded in 1975."

Nevertheless, offsetting, which has now become the preferred get-out clause, especially for the aviation sector, is no longer an option. It is cheap and focuses on forests, which are not a secure, long-term storage of carbon (see California or Australia). Therefore, what people really need are sustainable fuels to power aviation in the future, this will never happen if everybody is just buying cheap offsets.

8.7.4. Data for Action, Not Just Analysis

The old good saying "you can´t manage what you can't measure" has never been more valuable. But how so?

First, assess emissions and formulate a strategy to reduce footprint.

Second, set up a dashboard with key sustainability indicators.

Finally, be certain about your sustainability strengths when making prospective decisions: Can you foresee the impact of your next hotel opening? Is your next merge with a travel operator sustainable, i.e., positive in long-term economic standards?

8.7.5. Less Bottom-Line Operational Mindset, Circular Economy, and Sustainable Resource Management

From theory to harsh reality. Travel, tourism and hospitality (TTH) firms are increasingly aware of the importance of managing world´s limited resources. It is time to reduce wasteful consumption of resources as much as possible by reusing all available resources and minimizing waste.

8.7.6. Food Waste Is Not an Option (If It Ever Was)

One resource that requires special attention is food. Food has cultural, social, symbolic and heritage significance. As research shows, the food waste challenge can and must be solved with management practices and innovations.

8.7.7. Speed Up Business Action by Aligning with Sustainability Development Goals

Sustainability is not just climate change. Pay close attention to other UN Sustainable Development Goals (SDGs) and pledge to them. Several SDGs fall squarely into TTH company's core business principle: people service. For example, SDG8: "decent work and economic growth."

8.7.8. Be Proactive with Global Environmental Policy

Until recently, firms were compliant with local government regulations and policies. The road to sustainability has provided legitimacy to global players, such the UN Agenda (2030). TTH firms must get actively involved with these regulatory bodies and the new international global development agendas.

8.7.9. Put Environmental, Social, and Governance (ESG) In Your Corporate Agenda

Environmental, Social, and Governance (ESG) criteria are the three central factors considered when measuring the sustainability and societal impact of investment in a company or business. All the largest private equity investment funds have pledged to the Sustainability Standards and Policies laid out by the World Bank's International Finance Corporation.

8.7.10. Green-Washing Is Never An Option

It is time to answer those hard hitting, inconvenient questions: Inasmuch as airlines say otherwise, there is no sustainable cheap travel. Likewise, to this date, luxury accommodation and transportation are not sustainable businesses. Moreover, 2019 has seen a massive change in the perception of responsibility for the environment. Individuals make their choices more critically and expect companies to make significant contributions as well. Hence, in order to talk the talk is no longer enough – reputational risks are now very real. Therefore change must be seen as an opportunity. We have already supported many corporations to get a head start with clear benefits for early adopters implied.

8.8. TECHNOLOGIES ENABLING THE DIGITAL TOURISM ECONOMY

Technological advancements are having a deep impact on the tourism sector. These innovations range from business management technologies (e.g., mobile technologies/cloud computing, automation and advanced robotics, blockchain, data analytics, cloud computing), to technologies that produce innovative tourism products, services and experiences (e.g., virtual/augmented reality, Internet-of-Things), and technologies that assist, understand and connect with markets (e.g., data analytics, cloud computing, and artificial intelligence) (OECD, 2017c). Moreover, due to the information-intensive nature of tourism services, the breadth of information available on the internet, and ease of booking, tourism outperforms other sectors when it comes to the share of businesses making online sales in the 28 OECD countries for which data is available, with the exception of Canada, Finland and the United Kingdom.

OECD (2019) data show that on average 77% of the accommodation and food and beverage service businesses in OECD countries have a website or homepage and 70% use social media. The tourism sector has embraced e-commerce, as online platforms and payment systems have changed the way people buy travel products. Moreover, a report on electronic commerce (e-commerce) in the EU highlights that over 70% of internet users made at least one online purchase of goods and services over the previous 12-month period for private use. of that group, over half (54%) purchased travel and holiday accommodation, behind only clothes and sports goods, which were purchased by around two-thirds (65%). Hence, e-shoppers in the 25–54 age group were most likely to purchase travel and holidays (57%). The findings also indicated that the proportion of e-shoppers varied considerably across the EU, ranging from 29% in Romania, to 91% in the United Kingdom (Eurostat, 2020).

Figure 8.6. Technologies enabling the digital tourism economy.

Source: Image by unsplash.com.

However, while the digital transformation progressively touches all sectors in the economy, it does so with differing speeds and extents. Recent OECD work (Calvino et al., 2018) assesses the digital intensity of sectors by looking at the technological components of digitalization (tangible and intangible ICT investment, purchases of intermediate ICT goods and services, robots). Moreover, the human capital required to embed technology in production (ICT specialist intensity), and the ways in which digital technology impacts how firms' interface with the market (online sales). Hence, in a taxonomy of sectors by digital intensity, accommodation and food service activities is ranked low for digital intensity, while arts, entertainment and recreation activities are ranked medium-high based on seven different metrics (OECD, 2019c).

These results highlight that tourism businesses have significant untapped potential when it comes to the adoption and application of digital solutions, although they may benefit from ICT investments in other sectors through the outsourcing of certain business activities (e.g., sales and online advertising).

8.9. TOURISM TRENDS: THE LATEST OPPORTUNITIES FOR THE TOURISM INDUSTRY

Industries are continually disrupted by new trends and new innovations, and the tourism industry is no exception to this. Keeping pace with these emerging tourism trends can help businesses to stay competitive and cater to the needs of customers. Moreover, with COVID as a backdrop, some of these needs have evolved, priorities have shifted, and entirely new requirements have emerged. Read on to learn much more about the key trends within tourism for 2021. However, whatever aspect of the tourism sector your business is involved in, you need to keep your finger on the pulse. Formerly beloved concepts and products go out of style, to be replaced by more modern elements that end up capturing more of the market. Hence, new destinations, new technologies and means of transport have caused major shifts in the industry. Early adoption of new trends is vital.

8.9.1. General and COVID "Related" Tourism Trends

It is important to follow the current tourism trends in response to increasing consumer actions as a result of the coronary pandemic. However, most patterns have arisen from more general shifts in customer behavior. Below, you can see both general developments in hospitality, along with ways that serve as a solution to the coronavirus pandemic and associated shifts in consumer behavior.

8.9.2. Safety and Hygiene Tourism Trends

Whether it is airlines, cruises, hotels, restaurants or bars, since the outbreak of COVID, safety and hygiene standards have been paramount. Therefore, there are several tourism trends that are related to this, such as increased cleaning, socially distanced seating, providing hand gel and enforcing masks in some settings. This is also now a vital part of tourism marketing, with companies needing to make clear what their hygiene and safety policies are and what measures they are taking to keep customers safe. The threat of COVID has meant people are more reluctant to travel and visit tourism hot spots, so they will need to be persuaded that it is safe.

8.9.3. Increased Emphasis on Leisure

COVID has forced countries to adopt travel restrictions, while many businesses are encouraging employees to work from home and use video calling. As a result, business events have been particularly badly affected and one of the resulting tourism trends has been a switch in focus towards leisure customers. Moreover, the pandemic has been hard on people, and many are desperate for a holiday. If your business is typically focused on business customers, you may want to investigate ways to change this approach, at least temporarily. In the process, you will likely need to change your marketing messages and even the distribution channels you use to generate sales. Hence, depending on your business, you may wish to focus efforts on families, couples, or groups of friends, and you could potentially create package deals to appeal to these demographics. It is also a good idea to evaluate how your competitors have responded to the crisis and whether they are doing anything that you could take inspiration from.

8.9.4. Shift from International to Local

The various travel restrictions and the reluctance of many people to travel abroad has meant many in the tourism industry are having to focus on local customers, rather than international ones. This does not mean giving up on international travelers entirely, but it is likely to require a change in your core marketing strategies. With hotels, it could be best to highlight the kinds of facilities that may appeal to the local market, such as your restaurant, your gym facilities, your Wi-Fi and even the fact that your hotel rooms are ideal for remote work. Airlines and tourism management companies may also need to shift gears and prioritize domestic tourists. It is therefore worth remembering that local customers are less likely to cancel too, as they will only have to pay attention to local restrictions and are not as likely to have to quarantine after their visit.

8.9.5. Growth of Contactless Payments

Contactless payments have been a staple when it comes to technology in tourism for some time now, but the emergence of options like Google Pay and Apple Pay has helped to take this to the next level, meaning customers do not even need to carry around a debit card or credit card to pay for meals, hotel stays, transport, and other services. Moreover, allowing contactless payments has enabled tourism companies to reduce friction and improve the speed of check-ins and check-outs. It therefore means goods can be paid for swiftly, encouraging spontaneous purchases. With coronavirus, contactless payments are in greater demand than ever, as staff and customers often prefer to avoid handling cash.

8.10. GENERAL UPCOMING TOURISM TRENDS

There are a variety of tourism trends that are based on more general changes in consumer behavior, like the need for healthy and organic food & drinks, sustainability, personalized service the rising demand of digitalization and the use of technology. Below, you will be able to learn more about these more general tourism trends.

8.10.1. Voice Search and Voice Control

With home smart speakers growing in popularity, as well as mobile assistants like Siri, Google Assistant and Bixby, more and more tourism customers are turning to voice search. For those in the tourism industry, it is important to capture these guests by structuring website content properly, so that it appears in voice search and to allow for voice bookings. Hence, tourist information is a key part of the customer experience with many companies and voice control and AI can be invaluable here. Moreover, hotel rooms can include smart speakers or other IoT devices that are compatible with voice control, allowing users to more easily turn devices on and off, or change settings within their rooms.

8.10.2. Virtual Reality Tourism Trends

Virtual reality is another of the major tourism trends disrupting the industry and capitalizing on the technology can give you an edge over rivals who have not yet adopted it. Through online VR tours, customers can experience hotel interiors, restaurant interiors, outdoor tourist attractions and more, all from their home.

Crucially, they can do this at the decision-making phase of the customer journey. This can then be the difference between customers completing a booking or backing out and VR is especially useful within the context of COVID, where customers may have second thoughts and may need extra encouragement to press ahead with their plans. Moreover, most modern VR tours are also web-based, meaning they can be viewed through any mainstream web browser. The quality of the VR tour and the extent of immersion can then be improved further through VR headsets.

8.10.3. Solo Travel

Leisure travel used to be a family affair or something that couples undertook together. While that's still the case for many, more and more people are choosing to strike out on their own. Enjoying a solo trip is no longer so unusual and tourist trends increasingly reflect this. Therefore, the needs of solo travelers are diverse. Some, simply want to travel without the distraction of a companion. Others are young singles looking for social activities or to find a partner. Moreover, some widowed seniors even use long-term hotel stays or cruises as a luxurious alternative to conventional elder care. These tourism trends are set to grow and grow.

8.10.4. Eco Travel

Tourism trends are heavily influenced by the concerns and mores of the customer base. As a new generation becomes increasingly relevant in the marketplace, the ideals driving their purchasing decisions create new tourism trends. Eco travel is just one example of these tourism trends, reflecting a growing concern among today's travelers for ethical and sustainable tourism options. It therefore includes simple changes, such as the availability of carbon credits when booking a flight or the option to rent an electric instead of a conventional vehicle. More sophisticated examples might include tourism with a volunteer element, perhaps working on a nature reserve or engaging in conservation work.

8.10.5. Local Experience

Today's tourists do not want to be insulated from the places they visit inside a cultural bubble. They want to engage with and participate in the local culture. From enjoying local cuisine to celebrating regional festivals and holidays, local experiences are set to become some of the top tourist trends to watch.

One example of a popular local experience would be visiting Japan during a major festival, renting formal Japanese clothes to wear, consuming regional delicacies and engaging in traditional games or cultural activities. Another might be a long stay with a host family in the destination country to learn more about the local culture.

8.10.6. Robots, Chatbots, and Automation

One of the more eye-catching examples of these tourism trends is Connie, the Hilton Hotel chain's robot concierge. Other hotels have also got in on the robot-staff trend, installing interactive robots to handle certain reception duties or even having them serve food and drink to visitors. This kind of novelty application, however, is far from the only one. Many customers now book their travel and accommodation with the help of internet chatbots, specifically tailored AI who can handle queries and assist customers with useful information when human operators are unavailable.

8.11. CONCLUSION

In the conclusion, this chapter discussed about the significance of new business environment and trends in industry of tourism. This chapter also discussed about the role of digital technology in the sector of tourism. It discussed about the various new trends in tourism industry such as polar tourism, space tourism, dark tourism, voluntourism, luxury tourism, culinary tourism, and BRIC tourism. Moreover, it discussed about the importance of the marketing management trends in tourism and hospitality industry. This chapter also enlightens about the new microenvironment of tourism and hospitality industry. In this chapter, it has been explained that how to prepare tourism businesses for the digital future.

This chapter also discussed about the various sorts of sustainable tourism that have been playing significant role. Towards the end of the chapter, several different technologies that helps in enabling the digital tourism economy along with the upcoming tourism trends such as voice search and voice control, virtual reality tourism trends, solo travel, eco travel, local experience, and robots, chatbots and automation have been discussed.

REFERENCES

1. Carlos Martin-Rios, (2020). *Sustainable Tourism: 10 Most Important Trends in 2020.* [online] Hospitalityinsights.ehl.edu. Available at: https://hospitalityinsights.ehl.edu/sustainable-tourism-trends (accessed 8 August 2021).
2. Gustavo, N., (2013). Marketing Management Trends in Tourism and Hospitality Industry: Facing the 21st Century Environment. *International Journal of Marketing Studies,* [online] *5*(3). Available at: https://www.researchgate.net/publication/272805074_Marketing_Management_Trends_in_Tourism_and_Hospitality_Industry_Facing_the_21st_Century_Environment (accessed 8 August 2021).
3. OECD, (2020). Preparing tourism businesses for the digital future | OECD Tourism Trends and Policies 2020 | *OECD iLibrary*. [online] Oecd-ilibrary.org. Available at: https://www.oecd-ilibrary.org/sites/f528d444-en/index.html?itemId=/content/component/f528d444-en (accessed 8 August 2021).
4. Revfine, (2021). *Tourism Trends: 16 Opportunities for The Tourism Industry in 2021.* [online] Revfine.com. Available at: https://www.revfine.com/tourism-trends/ (accessed 8 August 2021).
5. Tutorials point, n.d. Tourism Management – Trends and Future – *Tutorialspoint.* [online] Tutorialspoint.com. Available at: https://www.tutorialspoint.com/tourism_management/tourism_management_trends_and_future.htm (accessed 8 August 2021).
6. Westcott and Anderson, W., (n.d.). *Top Trends*. [online] Opentextbc.ca. Available at: https://opentextbc.ca/introtourism2e/chapter/top-trends/ (accessed 8 August 2021).

INDEX

N

O

P

R

S

T